T0221094

COMPUT
CONTEXT

It were far better never to think of investigating truth at all, than to do so without a method . . . by a method I mean certain and simple rules, such that, if a man observe them accurately, he shall never assume what is false as true, and will never spend his mental efforts to no purpose.

Descartes

Against that positivism which stops before phenomena, saying "there are only facts," I should say: no, it is precisely facts that do not exist, only interpretations.

Nietzsche

COMPUTERS IN CONTEXT

The Philosophy and Practice of Systems Design

Bo Dahlbom
Lars Mathiassen

Blackwell
Publishing

© 1993 by Bo Dahlbom and Lars Mathiassen

BLACKWELL PUBLISHING
350 Main Street, Malden, MA 02148-5020, USA
108 Cowley Road, Oxford OX4 1JF, UK
550 Swanston Street, Carlton, Victoria 3053, Australia

The right of Bo Dahlbom and Lars Mathiassen to be identified as the Authors of
this Work has been asserted in accordance with the UK Copyright, Designs, and
Patents Act 1988.

First published 1995
Reprinted 1995, 1996, 1997, 1999, 2000 (twice), 2001, 2003, 2004

Library of Congress Cataloging-in-Publication Data

Dahlbom, Bo, 1949–
     Computers in context : the philosophy and practice of systems design /
Bo Dahlbom and Lars Mathiassen.
          p.   cm.
     Includes bibliographical references and index.
     ISBN 1-55786-430-6 (hbk) — ISBN 1-55786-405-5 (pbk)
     1. Systems design. 2. Computers and civilization.
     I. Mathiassen, Lars. II. Title.
QA76.9.588D25   1995   95–0779
004.2'1—dc20   CIP

A catalogue record for this title is available from the British Library.

Set in New Baskerville and Excelsior

For further information on
Blackwell Publishing, visit our website:
http://www.blackwellpublishing.com

Contents

Manifestly, no condition of life could be so well adapted for the practice of philosophy as this in which chance finds you today!

Marcus Aurelius

Preface

Computer professionals have to deal with a steady stream of methods and tools. With each new buzzword, there is a wave of conferences, books, and advertisements full of proposals and great promises. Entering the computer profession, we become excited by all this turbulence, but also quite overwhelmed. After a few years of trying to master all the trends, we begin to notice the repetitions, and we start wondering what the fuss is all about.

The most effective way to cope with the enormous amount of information is to ignore most of it. Rather than hurrying from one idea to the next, we should begin asking more fundamental questions. A less frantic reflection on our profession and its essential concepts, ideas, and challenges will help us see what is really new and will help us evaluate the trends in the light of the inherent problems involved in developing high-quality computer systems.

If we turn our back on the buzzwords for a while and reflect on our profession, what do we find? Beginning with the simple but at the same time most fundamental questions, we ask about systems development: What is it? How is it done? How is it done well? Why is it done? What major difficulties does it have to overcome? Such a reflection naturally divides into a discussion of *systems*, the things we work with, and the ways they influence our thinking; a discussion of *development*, the activity in which systems are being produced; a discussion of *quality*, the raison d'être of our profession; and a discussion of the fundamental contradictions that we have to deal with when actually getting down to *practice.* These are the ingredients we see in a philosophy of systems development. We have organized our book accordingly.

This book has been written for students and practitioners within the computing field, so we have avoided philosophical jargon and presuppose no acquaintance with philosophy. Our ambition has been not to turn computer professionals into philosophers but rather to encourage programmers, managers, and users to become more professional. We all acquire a philosophy in our education and practice, of course, but this book challenges its readers to make their philosophy explicit, to confront it with alternative perspectives and with their own practice. The goal is to take a few steps in the direction of a richer, clearer, more personal understanding of the practice of developing computer systems, in order to improve that practice.

This book is also intended for anyone concerned with computer technology and the relations between computers and society. The book does not presuppose a deep, technical understanding of computers and programming. For this audience, the book provides insights into the practices and problems involved in developing and using computer technology in organizations. To really understand how a technology is shaping society, we must understand how its engineers work and think.

The philosophical foundations of computer systems development are spelled out, but in a way that encourages readers to reflect upon what they already know, to work with their attitudes toward what they know and with their perspectives on their profession. We have tried to write a text that will open up the issues rather than close them by giving definite answers. We introduce a number of concepts, theories, and perspectives, but we also try to make readers curious; we try to surprise them and create opposition.

The book can be used both as a text in a college course and for individual study; for students in the computer or information sciences, for practicing designers of computer systems, and for anyone with an interest in technology and society; for graduate students as well as relatively advanced undergraduates; for students in computer engineering as well as those attending business schools. In order to make the book more accessible, we have chosen a less academic style, avoiding footnotes and references in the text. We have made up for this by including an extensive further reading section, in which we identify sources and do our best to acknowledge debts. To ease the burden of teaching a cross-disciplinary book such as this, we have included notes for the instructor at the end of the book, as well as student exercises and suggestions for further reading and useful supplementary texts for each chapter.

One of the authors of this book is a computer scientist in an engineering school, who is an amateur philosopher; the other is a philosopher from a department of technology and social change, who is acting chairman of an information systems department at a business school. A previous version of the book was used in computer and information science courses in Scandinavia. This revised version has very much benefited from comments and criticism by the students and teachers involved, as well as by numerous colleagues who very generously responded to our requests for feedback. Late in the process we have received good support from the editorial staff at Blackwell, and we are particularly grateful to our very professional editor, Susan Milmoe and to our careful copyeditor Stephanie Argeros-Magean.

We owe a special debt to the following friends and colleagues: Ivan Aaen, Liam Bannon, Gro Bjerknes, Lars Bækgaard, Susanne Bødker, Claudio Ciborra, Daniel Dennett, Lynda Davies, Rex Dumdum, Olle Edqvist, Pelle Ehn, Göran Goldkuhl, Joan Greenbaum, Jonathan Grudin, Johan Gåsemyr, Rudi Hirschheim, Kristo Ivanov, Malthe Jacobsen, Lars-Erik Janlert, Ingvar Johansson, Finn Kensing, Heinz Klein, Mikko Korpela, Börje Langefors, Giovan Francesco Lanzara, David Levinger, Kim Halskov Madsen, Michael Mandahl, Michael Manthey, Arve Meisingset, Peter Axel Nielsen, Torbjörn Nordström, Markku Nurminen, Horst Oberquelle, Agneta Olerup, Agneta Ranerup, Jan Stage,

Erik Stolterman, Carsten Sørensen, Kari Thoresen, Terry Winograd, and Heinz Züllighoven.

This book is truly a collaborative endeavor. Every para-
graph, every sentence is the result of discussions, of joint and
individual rewritings, of files being passed back and forth. It has
been natural to think of the production of the book as a systems
development project, beginning with the analysis and specification of requirements and going on to countless other tasks:
writing proposals for individual chapters, designing the book
top-down, writing and revising individual paragraphs, dividing
the paragraphs between us, revising our design in an iterative
fashion, going through the pain of eliminating sections of text,
worrying about version control, and so on. For both of us, the
writing of this book has been an enjoyable learning experience –
and we hope that the experience of reading the book will be a
similar one.

We are grateful to the Swedish Council for Planning and
Coordination of Research and to the Danish Natural Science
Research Council for economic support.

Göteborg and Aalborg, 1992

Bo Dahlbom
dahlbom@cs.chalmers.se

Lars Mathiassen
larsm@iesd.auc.dk

Part I

Systems

Computer professionals work with systems. They develop and maintain computer systems for others to use. And they use computer systems as their tools in doing this. No wonder they tend to see systems everywhere, no wonder they believe in systems thinking.

As computer systems have become vital and integral parts of organizational life and society at large, the computer professions have become increasingly numerous and powerful. This rapid expansion has made us optimistic and confident, even if in the reality of computer usage there is cause for a more critical and humble attitude.

We tend to forget that the idea of computing and systems thinking has a long heritage in our intellectual history. Looking closer at that heritage, studying the roots of our ideas, is a fruitful approach to a deeper understanding of the computer systems we work with today.

From the very beginning, in chapter 1, the computer is on stage as a solution looking for a problem. What is a computer? What are the fundamental ideas related to computing? How can computers be used? The computer is explained as a powerful realization of a mechanistic world view, stressing the idea of representation, the use of formalizations and rules, and a belief in rationality and bureaucratic organization.

If the computer is the solution, then information is the problem. In chapter 2, we discuss the use of information in organizations, and the problem of turning knowledge into information and information into data. This leads us to an alternative world view, romanticism, which stresses interpretation and change and thus expresses well the challenges we encounter when trying to put our computers to use.

In chapter 3, we turn from computer systems and information systems to the general idea of systems thinking. We discuss three frameworks for thinking about computers and information, and we spell out their different implications for systems development work.

1

Computers

Listen to Howard Aiken, physicist at Harvard and designer of some of the very first computers, speaking in 1956: "If it should ever turn out that the basic logics of a machine designed for the numerical solution of differential equations coincide with the logics of a machine intended to make bills for a department store, I would regard this as the most amazing coincidence that I have ever encountered."

Aiken was terribly mistaken, but certainly not for lack of computer technology expertise. Compare his prediction with that of John McCarthy, mathematician and inventor of Lisp, writing in that same year 1956 to the Rockefeller Foundation, asking for money to finance the first artificial intelligence conference at Dartmouth College: "The study is to proceed on the basis of the conjecture that every aspect of learning or any other feature of intelligence can in principle be so precisely described that a machine can be made to simulate it."

Aiken was very much influenced by the usage that had motivated him, and others, when constructing the first computing machines. McCarthy based *his* view of what computers could do on an understanding of the principles of a general computing machine, rather than on the actual capacity and use of the machines of his time. Together Aiken and McCarthy cover the field, if we are looking for answers to what computers can do.

Aiken is cautious and unimaginative; McCarthy is optimistic, at least when he goes from speaking in principle to predicting practice. But is he right in principle? What are the limits to the possible use of computers? And how are those limits set? By our imagination? By our needs? Or by the principles of computing?

Using Computers

Think of the first time you succeeded in making a computer do what you intended it to do. Think of the feeling it gave you to be in control of a powerful machine. Before getting too excited, don't forget the times you had to struggle to make the computer do what you wanted. You could not get the right result, or, worse, there was no reaction at all. Remember the direct way in which mistakes or inconsistencies in your first programs were revealed and the difficulties you sometimes had when you were trying to correct the programs? You knew that something was wrong, but you had no idea what it was.

Computers are fascinating because they are fast, powerful, and extremely versatile machines, and because they are programmable. They will obey our most whimsical commands, provided they can interpret them. It is really magic: They do exactly what we tell them to do. But, of course, we have to think and express ourselves clearly; we have to be careful with our words.

We can use computers to play with texts and immediately see the consequences. It is easy to reuse text and to experiment with different formulations and the sequence in which we present our argument. There is no guarantee, of course, that texts produced on computers are better in quality, or clearer and richer. Nevertheless, computers are extremely effective tools for producing, modifying, and combining texts, and they offer us fascinating opportunities to play and to experiment, while we try to be convincing, clear or even poetic.

We can also use computers to explore the world without having to suffer real-world consequences. Kids fight monsters without ever getting hurt. Pilots are trained in flight simulators without the hazards and high costs. Investments are evaluated without running the risk of losing fortunes. And bridges and highways are designed and tested without the risk of collapses or the inconvenience of traffic jams.

In general, we use computers to process, communicate, store, and keep track of information. Computers also provide us with new and useful opportunities, and they liberate us from many laborious and boring tasks.

It is not surprising that the development of computer systems raises difficult questions. We use computers to automate administrative tasks and to mechanize and automate production processes. But how can we make sure that the good qualities of the traditional manual way of doing things are not lost? And what about the people who lose their jobs because of the computer?

Computers provide us with data that give us a basis for decision making – but to what extent can we rely on those data? Are the data up-to-date and correct? What kinds of interpretations were made when the data were originally registered? What about the uncertainties introduced by our own interpretations?

Computers are used to monitor and control complex technical systems to minimize errors and avoid breakdowns and catastrophes. But computer systems are themselves complex artifacts that introduce new sources of error and uncertainty. And what about political issues? Do we want to use computers to keep track of people's every move and opinion? Do we want to use them to develop advanced military systems in outer space?

All these questions concern quality. People are concerned with the quality of work as computers replace old work habits and introduce new ones. We have to worry about the quality of data and information when computers are used, as they so often are, to provide decision support. As citizens we should be concerned with the quality of life in a society pervaded by information technology. And as systems developers we should be concerned about the quality of a particular computer system in relation to the wants and needs of the customer or user.

In this book, when we discuss the development and use of computer systems, we have all these aspects of quality, and more, in mind. In fact, our philosophy is that the development of computer systems is a constant struggle with quality. This struggle with quality constitutes the fundamental challenge and fascination of being a systems developer. Before expanding further on this theme, we must, however, go deeper into the very idea of computing.

The Mechanistic Heritage

Electronic computers have been around since the end of World War II. Before that time, computers were people working in big insurance companies or ballistic research laboratories performing long and tedious calculations. These human computers used desk calculators to perform simple subtasks of addition and multiplication, combining these subtasks into the computation of more complex functions.

During the war, new artillery weapons were developed at such a pace that the human computers were falling far behind in computing the necessary firing tables. As a result, the U.S. government was interested in supporting attempts to construct "an automatic calculator." Early machines such as Aiken's Mark 1 used electromagnetic relay technology. The decisive step to a full-blown computing machine was taken when the machines were made electronic, and with the capability of storing programs in their memories. The electronic representation of data made it possible to change the contents of registers much faster than in the mechanical machines. The idea of looking at computational procedures as data and of storing programs in memory made it possible and easy to change the function to be computed.

Electronic computers were built to replace human computers. They were designed as technical devices to be fed with numbers and computational procedures. They could compute according to prescribed procedures and deliver the computed results as output.

Classical computer science has linked human computing and machine computing in the so-called Church-Turing thesis: Everything that a human being can compute can be computed by a machine. According to this thesis, our intuitive understanding of computation refers to the set of computations that can be formally prescribed. Closely related to this idea is the concept of an algorithm.

An *algorithm* is an instruction (or a plan or procedure) indicating how to manipulate some given input to produce a desired output. An instruction is an algorithm if it is finite, definite, and effective. That it is finite means that it can be mathematically proved to terminate – that is, to finish after a finite number of steps. That it is definite means that each manipulation is well defined and precise. That it is effective means that each manipulation is so elementary that it can be performed by a human being using just pen and paper, or by a machine.

The algorithm, with its strict definition, is a bridge between our intuitive notion of (human) computation and the idea of a computing machine. If human computation is algorithmic, then, since any algorithm can be implemented on a machine, the Church-Turing thesis is true. But notice that this thesis cannot be mathematically proved, since it relies on an intuition about what it is for a human being to compute.

The construction of the electronic computer was the crowning achievement of a long tradition in our culture that identified thinking with computation. This theory of thinking was clearly formulated already in the seventeenth century by the French philosopher and scientist René Descartes (1596–1650). Inspired by his work in mathematics, Descartes developed a theory of thinking as the rational manipulation of symbols by means of rules. When people spoke of the early computers as electronic brains, as intelligent, thinking machines, they simply applied the Cartesian idea of thinking as computation.

But something is obviously wrong here. Thinking is the essence of being human, according to Descartes, the human mind is a thinking thing, a *res cogitans*, as he puts it. But the computers we see around us are anything but human. They don't reason, argue, plan, fantasize, imagine, memorize, hope, or foresee. They compute but they don't think. Does this mean that Descartes was wrong? And not only Descartes, but our whole modern conception of thinking underlying the construction of the computer and the development of computer science? Some would say yes, there is something wrong here. Others would argue that it is only a matter of programming: If the computer on your desk does not seem to be a thinking thing, this is only because there has not been enough programming done.

Like so many other questions raised by the ongoing comput-erization of work and society, our answer to this question will express our fundamental view of the world. If our answer is "yes, with more programming, computers will become thinking ma-chines," it is likely that we will agree with many of the ideas of the *mechanistic* world view as developed by the great seventeenth century system builders, like Descartes and the German philoso-pher and mathematician Gottfried Wilhelm von Leibniz (1646 – 1716). In this world view we find a general and powerful idea about computation, related to ideas of representation, formaliza-tion, program, order, and control.

If our answer is "no, a machine can never be made to think," then we are probably attuned to the *romantic* world view of the early nineteenth century. This is a very different conception of the world based on ideas about interpretation, uniqueness, chaos, and change. In the next chapter we shall discuss this romantic alternative.

We shall begin in this this chapter by taking a more careful look at our mechanistic heritage. This heritage exerts a powerful influence on all of us, whether we believe in the project of turning computers into thinking machines or not.

But before we do that, let's discuss world views. We don't often find reason to think deeply about how we view the world, and we seldom have to formulate or pledge allegiance to a coherent world view. Our world views tend to consist of more or less loosely related, sometimes conflicting elements, some explicit and some tacit, gathered from very different sources.

Most of us have world views that combine both mechanistic and romantic ideas, even if we would not normally identify them by these terms. The mere fact that you are interested in computers means you are mechanistically oriented. And in order to be the kind of person who would read a book like this, you probably have to be touched by romantic ideas.

A discussion of the mechanistic world view is not only an examination of ideas that have dominated the development and use of the electronic computer. It is also a way of reflecting on our own ideas about computers and their use, reflecting on the roots, nature, and legitimacy of those ideas.

Representation and Formalization

A powerful process of change, called modernization, began in sixteenth-century Europe. The modernization process affected all aspects of human life. It changed a society dominated by farming and craft, religion and miracles, authority and tradition into a world of technology and science, democracy and liberty, progress and revolutions. Religious upheavals, industrial expansion, the central perspective in art, the exploitation of America and the invention of the machine were all parts of this process of modernization.

To come to grips with their changing world, the modern Europeans developed natural science. The new physics taught that our naive perception of the world is loaded with errors, that the world is not what it seems to be. The sun seems to be moving across the sky, but it isn't. Objects seem to be colored, but they aren't. With the recognition of systematic perceptual errors, the world and our representation of it become clearly distinguished.

Our perception of the world does not coincide with the world itself. The world itself is out there; our experience of it is in here, in the mind.

This dualism of external world and inner life that we now take for granted is really a modern idea. The explicit distinction between the world and our representations of it was a necessary condition for the project of modern science: to replace our naive perceptions with true, scientifically based representations of reality.

No one did more than Descartes to promote the idea of knowledge as the representation in the mind of a world out there. The mind is a mirror of the world, Descartes could say in a time when glassmaking had advanced enough to make mirrors high fashion. But an imperfect, cloudy mirror distorts its object, and our minds are not to be trusted, so we have to make sure that our ideas are clear and exact before we depend on them to give us an accurate picture of the world.

Unclear ideas confuse us about the world and get confused with one another. Images, the pictures of the world we see in our mind, contain an abundance of material – subjective properties like color, taste, and smell – that is not to be found in the world. The mechanistic world view challenges us to strip our ideas of such subjective material in order to make them true to the world.

Like Galileo (1564–1642) before him, Descartes came to argue that we have to use mathematics as a means of representation in order to map the world in a clear and exact way. The world is like an open book for us to read, and it is written in the language of mathematics.

Galileo worked with geometry, using lines, triangles, and squares to represent properties like velocity and acceleration. Transforming geometry into algebra in his analytic geometry, Descartes could do mechanics with numbers and algebraic functions and argue that algebra provides us with the exact symbols we need to represent the world truthfully.

Today, it is easy for us to recognize the powerful idea of formalization in the mechanistic struggle to arrive at clear and exact ideas. But the mechanists wanted to formalize not only the ideas or symbols we use to represent the world, but also the process of thinking itself. With the use of mathematical symbolism as a means of representation follows the idea that thinking is the manipulation of these symbols, that thinking is mathematical reasoning, calculation, and proof. What distinguishes mathematical reasoning from everyday pondering is its explicit reliance on rules, on logic. Just as the use of mathematics in representing the world is motivated by a desire for exact, explicit knowledge, so the idea of thinking as computation is developed to give us a conception of thinking as exact and explicit rule following.

We have to realize, of course, that the idea of thinking as mathematical computation, as formulated by Descartes and later developed by Leibniz, was more of an idea than a full-blown theory. Both symbolism and mathematical reasoning were in those days still fairly informal and needed much work before they could be truly described as exact and explicit.

Leibniz worked hard on developing a mathematical notation, a universal calculus, that could serve as means for representing and reasoning about all our knowledge of the world. As part of this project he built several small calculators to test his ideas of exact reasoning. In these efforts he was motivated by the modern belief in knowledge as the universal problem solver and means to progress. He hoped that such a language, used both locally and in international diplomacy, would put an end to conflicts of all kinds, based as they were on misunderstanding due to the use of inexact, informal language. Indeed, this was a powerful dream in the seventeenth century, in a Europe ravaged by wars and religious conflicts.

By introducing explicit rules that tell us how to express ourselves, we can develop more exact representations and avoid misunderstanding. The computer, as we have come to know it, is based on the application of explicit rules and the idea of representation.

For computers to be of any use, we have to agree on how to apply a certain concept or how to interpret a specific physical state as a symbol. The computer is of no use without formalizations. And without the computer there would be one reason less to formalize observations about patients in hospitals, or information about employees and customers in the organizations where we work.

The mechanistic ideas of representation and formalization are at the very heart of computing. Data are representations of facts, and the computer is a technology for storing and manipulating data. Without the idea of representation, we would have no computers. Without computers, we would not have the same potential for automatically manipulating our representations of the world.

Rules and Rationality

The modern world view is a rational view of the world. Rational thinking is the conscious, competent administration of ideas, aided by a method. To rationalize is to rely on rules, to develop methods, write up programs. To be truly rational, we need not only to follow rules but also to know and be able to state and defend the rules we are following in our thinking. Before we undertake an action, we formulate the rules; before we develop a system, we formulate our method. The real work lies in choosing, formulating, and motivating the rules, the method. The rest is routine. A machine can do it.

Modern science with its method is only one example of a social institution rationalized by explicit rules. Descartes' scientific method – his "rules for the direction of the mind" – has its counterpart in the explicit and unequivocal rules controlling a bureaucracy, in the written instruction manual for operating and repairing a modern machine, in the explicitly formulated constitution of a modern democratic society, in the official curriculum for a modern educational institution, and so on.

For the mechanists, it is the business of science to map the world, to give a systematic, preferably axiomatized, definite, and

sys·tem (sĭs′təm), *n.* **1.** an assemblage or combination of things or parts forming a complex or unitary whole: *a mountain system, a railroad system.* **2.** any assemblage or set of correlated members: *a system of currency, a system of shorthand characters.* **3.** an ordered and comprehensive assemblage of facts, principles, doctrines, or the like in a particular field of knowledge or thought: *a system of philosophy.* **4.** a coördinated body of methods, or a complex scheme or plan of procedure: *a system of government, a penal system.* **5.** any formulated, regular, or special method or plan of procedure: *a system of marking, numbering, or measuring.* **6.** due method, or orderly manner of arrangement or procedure: *have system in one's work.* **7.** a number of heavenly bodies associated and acting together according to certain natural laws: *the solar system.* **8.** the world or universe. **9.** *Astron.* a hypothesis or theory of the disposition and arrangements of the heavenly bodies by which their phenomena, motions, changes, etc., are explained: *the Ptolemaic system, the Copernican system.* **10.** *Biol.* **a.** an assemblage of parts of organs of the same or similar tissues, or concerned with the same function: *the nervous system, the digestive system.* **b.** the entire human or animal body: *to expel poison from the system.* **11.** a method or scheme of classification: *the Linnean system of plants.* **12.** *Geol.* a major division of rocks comprising sedimentary deposits and igneous masses formed during a geological period. **13.** *Phys. Chem.* **a.** any substance or group of substances considered apart from the surroundings. **b.** a sample of matter consisting of one or more components in equilibrium in one or more phases. A system is called binary if containing two components; ternary, if containing three, etc. [t. LL: m. *systēma*, t. Gk.: organized whole] —**sys′tem·less,** *adj.*

true account of the world. This is possible because the world itself is an ordered, fundamentally unchanging system.

Newton's mechanics, with its three basic laws, is the outstanding example of an ordered system. Leibniz gave us a particularly powerful conception of the world as such a system, a world that is deterministic and governed by two fundamental principles: the principle of sufficient reason and the principle of preestablished harmony. Nothing can exist without a reason, and everything that exists has to be in harmony with everything else that exists.

The seventeenth-century intellectuals were fascinated by machines, and this interest influenced their image of the world. Descartes and Leibniz both played with machines as toys and as powerful ideas. They saw not only the world as a machine, but also the body (Descartes) and society (Leibniz). They laid the foundation, as we have seen, for viewing thinking itself as a mechanical process.

When we see something as a machine, be it the world, the body, or society, we want to take it apart and to make explicit the rules governing its behavior. When we know the rules, or perhaps better the laws, governing the functioning of a machine, then we can control it. When we realize that the heart is a pump, we know how to deal with it. A society governed by rules can be controlled by those formulating the rules. To follow rules in our thinking is to control our thinking.

The mechanistic emphasis on rationalism, on methods and programs, has had a strong impact on the development of computer technology and on the ways we think of programming and systems development.

To use computers we need programs, and to program we need methods. The history of computing can be seen as a continuing attempt to develop programming languages and methods for programming.

But the real challenge in computing lies in exploring and applying the mechanistic world view while at the same time understanding and appreciating the limits of its application.

When we try to control the world with computer programs or methods for systems development, we should not forget that the mechanistic world view is based on the assumption that the world we are trying to understand and control is itself an ordered, fundamentally unchanging system.

Computers and Bureaucracies

The mechanistic world view has influenced organizations and society long before the invention of the electronic computer, and we can learn a lot about computers and formalization by looking more generally at the ways we organize human activities.

Every organized human activity – from producing milk to developing computer systems – gives rise to two fundamental and opposite requirements: the division of labor into various tasks and the coordination of these tasks to accomplish the activity as a whole.

If more than one person participates in a systems development project, we must divide the task by designing the system as a set of related modules. Such modules are not simply the parts of the final system. They also define separate work tasks to be performed as part of the development effort. One of the important functions of a design specification is to serve as a basis for the division of labor during a project.

But a good design document, dividing the labor well, does not automatically lead to a satisfactory system. During implementation, we must evaluate and test the individual modules and, even more crucial, perform integration tests and evaluate the operation of the total system. In addition to design documents defining the division of labor, we need other techniques to support the coordination of individual tasks to accomplish a satisfactory result as a whole.

Organizations use different strategies for dividing labor and achieving coordination between individuals and groups. Shipyards, textile factories, hospitals, car repair shops and systems development projects each have different approaches to effective organization.

To understand the differences between organizations and to facilitate the design of effective organizations, we identify a number of abstract or ideal types of organization with different structural characteristics. The most well-known is the bureaucracy.

An organization is *bureaucratic* to the extent that the behavior of its actors is predetermined or predictable. The bureaucratic approach to organization relies on rules in prescribing behavior and achieving coordination. Bureaucracies are programmed. The assumption is that we know in advance what to do: The task uncertainty of the organization is low.

Bureaucracies are designed to be efficient by minimizing direct interaction between individuals and groups. Coordination is achieved by having each group or individual follow prescribed rules. When the rules do not apply, decisions are made by a hierarchy of managers, the hierarchy being the most efficient way to organize communication.

In a bureaucracy, management is kept separate from actual production. Workers are not supposed to make decisions. They produce goods or services according to instructions, only informing their managers about deviations and problems. Managers make decisions. They develop new plans and formulate instructions based on previous plans and status information.

A bureaucracy is like a computer, and like the computer, it is a powerful expression of mechanistic ideals. A bureaucratic organization is programmed, its work tasks are explicitly defined and formalized. It is a machine in which computing machines have their natural place, providing efficient processing and communication of information about products, activities, and resources.

The computer is a perfect bureaucrat, and it invites us to think like bureaucrats. We cannot use it without formalizing. The formalization imperative is in most cases quite obvious. We cannot develop a computer-based accounting system without formalizing what is meant by an account and what is meant by various transactions on an account. In other cases the formalization imperative is less obvious even if it is still there. In developing a computer-based text processing system, we do not have to formalize what we write about. But we do have to formalize the format in which we write it. Otherwise the computer will be of little help in manipulating, storing, and communicating the text.

Traditional production control systems provide a classic example of the bureaucratic use of computers. Coordination is viewed as a rational decision process where status information is produced on the shop floor and compiled through the computer system. Plans are created by production managers and supervisors on different levels of detail and these plans are distributed through the computer system.

The role of the supervisor is to make detailed plans expressing what each individual worker has to do and how the machines on the shop floor should be utilized. These plans are communicated to the workers, and they report back to their supervisors when jobs are finished or breakdowns occur. The supervisors make decisions on how to handle delays and breakdowns, compile reports to the production planner, and receive overall plans for the production in their departments.

The computer is used to communicate, process, and store the information needed to manage and coordinate the activities on the shop floor into one integrated effort. Workers, supervisors, and production planners use the computer to generate, receive, and process information. The assumption is that the information provided by the computer system represents the actual state of production and commitments. The computer is used in accordance with the mechanistic view of the world to support rational decision making in optimizing the management of resources.

In our everyday activities we all rely on bureaucratic approaches – and we have begun to use computers to support us in doing so. Yet we tend to forget the basic weakness of the bureaucratic approach. When the environment changes and the task uncertainty increases, we are ill-prepared. Since our behavior and thinking have been shaped by bureaucratic procedures, we are unwilling to engage in change, and we don't have the necessary resources to do so. Our computers are not making it any easier.

Coping with Change

An organization operates in an *environment*. This environment comprises everything outside the control of the organization that is of importance to its performance. This includes the nature of its products, customers, and competitors, the economic and political climate in which it must operate, and the kind of technologies on which it depends.

The environment of a specific organization may range from stable to dynamic, from that of the stonemason whose customers demand the same product decade after decade, to that of a medical research team trying to develop a new and effective medicine. A stable environment may, of course, change over time, but the variations are predictable. In contrast, the environment of the research team is dynamic in the sense of being highly unpredictable. The team has one or more theories on how to cure a specific disease, but they have no certain knowledge about what technologies to use, nor can they effectively predict whether and when they will succeed in their efforts. They simply have to try.

The effectiveness of an organizational structure is strongly dependent on the environment of the organization. The economic use of information is the strength, but also the weakness, of the bureaucracy. The more the environment of a bureaucracy changes, the more often its rules will not apply, and the more the management hierarchy will have to intervene. The Achilles heel of a bureaucracy is its inability to respond effectively to change.

When a bureaucratic organization fails, more organic structures will emerge. An *organic* structure can be defined by the absence of formalization, by the absence of explicit rules prescribing or determining the behavior of the actors. Bureaucratic structures are hierarchical; organic structures are more like networks.

The organic approach to organization relies on informal and direct interaction to achieve coordination between individuals and groups. The assumption is that the task uncertainty of the organization is high. New information related to organizing the activity will therefore become available as the activity is performed. To make this information useful, the involved actors must have the opportunity and obligation to communicate and interact. All actors are supposed to engage actively in decision making and planning as the activities are performed. Management and production are integrated.

Bureaucratic organizations are like machines, organic organizations like living organisms. Bureaucracies are based on a belief in and a striving for stability, order, and control. Organic structures are designed to cope effectively with dynamic environments. They go beyond the basic assumptions of the mechanistic world view, offering a constructive response to the limitations of bureaucracies.

Strangely enough, the computer has proved to be a useful element in organic strategies. In electronic mail systems, some degree of formalization is required. Communication forms and facilities for storing and retrieving information are formalized. Such systems also have to provide formal information about the addresses in the network. The success of electronic mail systems, however, lies in the informal and direct way in which communication is possible.

Electronic mail systems are explicitly designed to support organic coordination. But traditional computer systems can also be reinterpreted from an organic point of view. Let's see how we can learn more about the use of a traditional production control system by such a reinterpretation.

From an organic perspective, coordination can be viewed as a dynamic negotiation and creation of commitments. The role of the supervisor is to act as an intermediary between the production planners and the workers. As in the bureaucracy, workers, supervisors, and production planners communicate in a cooperative effort to coordinate activities. But at the same time they play opportunistic games for individual gain. Information is not necessarily to be trusted, because individual actors might withhold information or, even worse, generate misinformation to gain a personal advantage.

Supervisors are not primarily decision makers. They have to listen, negotiate, and manipulate in order to manage resources in a satisfactory way. As in the bureaucratic case, the computer-based production control system is used to communicate, process, and store information. But to be effective, the computer system should also support the actors in negotiating and administrating commitments, in addition to traditional production planning.

The organization is no longer viewed as a machine but rather as a dynamic network or an organism. Uncertainty of information has become an important issue and the best we can hope for is a satisfactory, rather than optimal, utilization of resources on the shop floor.

We often rely on organic approaches to organization in our everyday activities. We discuss with colleagues, we participate in ad hoc meetings, and we are members of autonomous groups that have been assigned more or less well-defined tasks. We have also become accustomed to using computers to do all this.

People cooperate across organizational and national boundaries. Project groups are formed in which the members are organizationally and geographically dispersed. These kinds of organic structures would not be effective, or even possible, without technologies like the telephone that can link the actors together in an informal and highly interactive fashion. But computer networks and hypermedia radically increase our ability to informally and directly coordinate activities between groups and individual actors across organizational and geographical boundaries.

Before the picture we have painted of organic forms of thinking and organizing becomes too glowing, we had better stop and consider its weaknesses. It then becomes clear, of course, that with an organic approach we do not utilize the fundamental strength of the computer: its ability to process formalized information. When using a computer, we do not have to write each letter from scratch. We can use a standardized format, and let the computer aid us in comparing, categorizing, and evaluating messages. Often it is convenient to have some kind of standardized format, or even some standardized ways of providing the information by enumerating in advance the possible choices, as in a traditional questionnaire.

There are both practical and economic reasons for formalizing information when using computers. In general, we know that the more stable and repetitive our work, the more we tend to apply bureaucratic approaches – not only because we are told to do so but because we find it effective and helpful.

As exemplified here, we can go back and forth between the way we think about computers and the way we think about social organizations, and in the process learn more about both. Mechanistic ideals and the formalization imperative continue to have a strong influence on both computers and organizations. Modular thinking and hierarchical structures have dominated our conception of computer systems. To deal with the complexity of these systems, we have struggled to separate concerns as we divide labor in the classical bureaucracy. We have also tried to economize with exchange of information by relying on hierarchical structures, thereby minimizing the number of active information channels between modules. We have been very successful in turning the computer into a perfect bureaucrat and computer systems into well-ordered bureaucracies.

But as a consequence of this, traditional computer technology is not particularly well suited to more organic organizations. And its bureaucratic nature has been a restraint on our imagination in finding new use for the technology. For a long time, the

use of computer technology was mainly restricted to the bureaucratic functions of organizations and it acted as a support for the
bureaucratic nature of those organizations. But that is changing
now as our interest in organic forms of organization has grown
enough to make us begin to rethink the very nature of computer
technology. Parallel architectures, neural nets, networks, and
hyperstructures are all examples of more organic ways of thinking of computers and computer systems. New ways of organizing
have changed the way we think about computers and vice versa.

The computer artifact has long since outgrown the image
of the human computer. As its use has diversified, our understanding of it has diversified as well. If we are working with
electronic mail systems, we naturally focus on those qualities of
computers that make us think of them as media. If we are
working with statistical program packages, we will attend to other
qualities and see the artifact as a computing machine. If we are
evaluating a text processing system, it is best to treat the system as
a tool for producing texts. But in many cases several competing
perspectives are equally relevant in understanding and designing
computer systems. If all you have is a hammer, the world will look
like a nail. It isn't.

2

Information

Between Data and Practice
Making Knowledge Explicit
Platonic and Aristotelian Concepts
The Romantic Challenge
A Dialectic Synthesis

It is the business of systems developers to apply computer technology in designing better information systems for organizations. As the use of computer technology has diversified, traditional information systems have been complemented with a rich variety of different systems. In that process, the task of the systems developer has changed as well, of course. Still, the computer technology remains an information technology. In chapter 1 our subject was the computer. Here we shall discuss its use by focusing on information.

Information is challenging. We all have to decide what to do with our lives. We make choices, and in order to choose what is best for us, we need information. No wonder information is highly valued. Yet, we all know that in our daily lives we are unbelievably careless both in gathering information and in making decisions. We take gossip seriously, listen only to what we want to hear, make snap decisions, and invent excuses to postpone making a choice.

Believe it or not, organizations are no better than individuals in this respect. Organizations put a high value on information and are willing to spend much time and money on it. Often, however, information is requested and collected only to be forgotten when decisions are made. Also, information is gathered that has little decision relevance, and it is often collected when the decision has already been made.

Organizations are constantly asking for more information while at the same time they are not using the information they have in a rational way. The situation is obviously not satisfactory. But what is the remedy? Should we conclude that we need a better technology to handle information for us? Or, should we conclude that the value of information is overrated?

Perhaps our frantic search for information serves the purpose of making us *seem* rational rather than being a real basis for rational decision making? Perhaps information gathering is an excuse for inaction rather than a means for rational action? Perhaps investments in information technology will be only another way of pretending to be rational, another excuse for not engaging in rational decision-making.

Even if this sorry picture of how we misuse and avoid using information is correct, there is no way around the platitude that information is valuable. And since it is valuable we must accept the challenge to consciously design our use of information, if not in our daily lives, at least in our organizations. Rather than making us depressed, the sorry picture painted above will indicate what a formidable task we have set ourselves as systems developers.

If organizations are irrational in their use of information, they will be irrational in their use of computer technology as well. Our task is not simply to design good information systems for a fictitious, ideally rational organization, but to design information systems for real, irrational organizations. In order to do this, it is not enough to simply define information as a resource that it is our task to organize the supply of. We must also understand the multiple roles played by information in organizations.

As systems developers we are ourselves producing, acquiring, and keeping track of lots of information. We use this information to organize the development effort and to provide other actors with platforms for making decisions. The very nature of our task is such that we have to deal professionally with information and its use.

Between Data and Practice

What is *information?* Answering this question is certainly not easy, and we are not much better off if we seek help in dictionaries. Information is somehow related to knowledge, but it is not the same as knowledge. More loosely, information is related to teaching, briefing, decision making, persuasion, manipulation, and even hearsay.

We all use and produce information. We ask for information and receive information in return. We receive lots of information without asking. There are information offices and information services. Still, when having to explain what information really is we remain vague: Information is something that gives knowledge, something that is related to the communication, transmission, or dissemination of knowledge.

Even if we cannot define information, we know very well when we have it and when we don't. As intensive consumers, managers, and purveyors of information, we are concerned with quality of information, and we use a variety of criteria to evaluate information.

We want information to be relevant. The information we use should address matters we are concerned with and preferably help us understand, make decisions, or act. We also want information to be reliable. The information should be true to the actual state of affairs. If we are told that a friend will arrive on the next train, we expect that she will actually do so. Sometimes we prefer information without redundancy – that is, without repetition. In other cases, we prefer some redundancy because this makes it easier to interpret or grasp the information. In addition, we are concerned with consistency, stability, comprehensibility, availability, retrievability, and other qualities of information.

Data, information, knowledge, and competence correspond to different levels or forms of human activity. In trying to be more precise about what information is, we can inquire further into the relations between information and these other levels of human activity. On the one hand we can ask: How does information relate to data? On the other hand we can ask: How does information relate to knowledge, competence, and human practice?

Data are a formalized representation of information, making it possible to process or communicate that information. Information is not the same as data. Assume that you receive a written message that reads "Call your mother." This piece of paper contains data. The signs on the paper are a representation of information, that someone wants you to perform a specific action. It is a formalized representation written according to well-known rules and conventions for using elementary signs and for combining these into more complex signs. Moreover, this representation makes further processing possible. The people – or machines – who know, or have access to, the relevant rules and conventions can read the signs and communicate them to others.

There are important differences that make it possible for us to distinguish between data and information. But still, hard questions come easy when we inquire further into the relation between information and data. Can we process and communicate information without using data? Can data exist independently of a human interpreter? How do we distinguish formal from informal expressions?

Turning to the broader context, we can look at the relation between information, on the one hand, and knowledge, competence, and *human practice* on the other. The concept of information is close to the concepts of knowledge and competence, but it also involves the concepts of interpreting and making ideas explicit. To produce information, we have to interpret what we experience and make explicit what we know. By doing so, we create the opportunity for others to share with us what we see, what we want, what we know, or what we believe.

Information comes in bits and pieces; knowledge and competence do not. Information is explicitly expressed in the form of signs and externally materialized as sound, print on paper, or electronically lit pixels on a screen. In contrast, knowledge and competence are personal and intrinsically related to each individual's practice. Information is something we provide and receive, knowledge and competence are something we have.

Again, there are important differences that make it possible for us to distinguish between information, on the one hand, and knowledge, competence, and human practice on the other. Still, as before, hard questions come easy when we inquire further into this relation. Can information exist independently of its producer or consumer? Are there certain kinds of information, which we might call facts, that do not require interpretation? To what extent can we and should we try to make knowledge and competence explicit?

Data, information, knowledge, and competence are closely related concepts, and the boundaries between them are fuzzy. The concept of information is, however, distinguishable from the others. To understand these differences, we have to look more closely at the ideas of interpreting experience, making knowledge explicit, and formalizing information.

The aim of a typical systems development project is to support or replace a human practice. So the first step involves learning about that practice. In doing so, systems developers are like anthropologists trying to understand and interpret the practices of a foreign culture.

If the practice in question is in a field with which we are unfamiliar, we will have to find reliable and representative informants that are willing to spend time sharing their knowledge with us. The more unfamiliar we are with the field, the greater our problems will be in securing cooperation, understanding what is being said, and deciding when our task is finished.

How do we choose an informant? Or how do we handle the situation when we begin suspecting that the informant that has been assigned to us is not really knowledgeable enough? At the very least, we need to be very good at communicating with experts in a field in which we have no expertise.

In anthropology, the discussion of methods for securing knowledge about alien cultures is an intense one. Is it really possible for us to understand the practices of an alien culture, we ask. Do we have to go native to do that and, if so, how do we then translate our observations into the vernacular of Western social science? Even if most systems development clients are not as alien as Trobrianders to us, we share with anthropology the problems of understanding their practice.

In another respect we are even worse off. Anthropology field workers can often define their task as that of charting a culture at a definite time, leaving questions of how the culture develops for others to worry about. When transforming the knowledge of human beings into information to be used in an information system, the idea is to use that system in a dynamic organization or in a rapidly changing, competitive market. As systems designers, we cannot turn our backs on the problem of change. An information system can become obsolete even before installation and will, if it is used, become a hindrance rather than a support for the organization.

One of the lessons of anthropology is that a culture should be observed in practice, preferably by participating in that practice. Our knowledge is interwoven with our practice, and that practice is normally dependent on a definite situation. This is clearly true for the knowledge we use every day. It is easy to get to work, but it is often difficult to give exact instructions about how to do it. When you are in your office, you can usually find the paper you are looking for, but try directing someone else in the same task over the phone. When you ask a psychologist what psychology is, she will hem and haw, but let her show you one of her experiments and you will see an example of confident knowledge in action.

Philosophers have traditionally located knowledge in the soul or mind. Our knowledge of the world is a map that we carry in our head. Transferring such a map to a computer should not be all that difficult. But if our knowledge is in our practice, in the situations where we perform that practice, our task becomes more uncertain. What are we then as systems developers supposed to do?

If we decide to limit ourselves to charting the knowledge we have in our heads, we will end up with a system that is dependent on a situated practice that it does not possess. To arrive at a richer system, we must somehow turn practices and situations into information. We all know how to do this, since we are used to talking about what we do. But the point is that we can only do it by relying on shared practices and experiences of situations. Think of what a cookbook for a true novice would look like. Every recipe would begin: "Turn on the light in the kitchen."

Making Knowledge Explicit

When we want to turn someone's knowledge into information and information into data, we will often have to make explicit what has hitherto been tacit in the competence of the individual having the knowledge. This can be a difficult task fraught with many obstacles.

The act of writing is one way of making things explicit. The idea that words can be used to make thinking explicit is a modern idea, dating back to the fascination created by the printed book when it began to be widely circulated in the late sixteenth century. The ideal of explicitness was then formulated in the slogan "like an open book." Once you realize, Galileo said, that mathematics is the true language of science, nature will become like an open book. All you have to do is read it. This conception of the language of mathematics, later developed by philosophers like Descartes and Leibniz, is that of an ideal language that can be read without interpretation. There can be no disagreement about what is being said, when it is said in an ideal language.

When Gutenberg invented the printing press in 1453, he not only made books available on a mass market but also

"I just can't tell from here. . . . That could either be our flock, another flock, or just a bunch of little m's"

changed our ideas about thinking and language. After Gutenberg, the printed word became a powerful social force. Education changed its character from training, under the personal supervision of a master, to the reading of books. The rhetoric of politics turned from speeches to pamphlets. Having information, being informed, began displacing having wisdom, being wise, as an ideal of knowledge. Knowledge began to be collected in books and the French eighteenth-century philosophers began compiling the first anonymous *Encyclopaedia*, supposed to contain all knowledge that was worth knowing.

Before Gutenberg, people agreed with Plato's derogatory view of writing as a bleak and rudimentary copy of thinking and oral communication: "Then anyone who leaves behind him a written manual, and likewise anyone who takes it over from him, on the supposition that such writing will provide something reliable and permanent, must be exceedingly simple-minded . . . if he imagines that written words can do anything more than remind one who knows that which the writing is concerned with." To Plato, the writing of books was a craft that was clearly inferior to the practice of knowledge conveyed by thinking and oral discussion.

When people began reading books, they read the Bible. But that is not an easy book to read. Adam and Eve begot Cain and Abel. These were the first people on earth. Cain slew Abel and then went to a foreign country and took himself a woman. Where did she come from? In order to understand the Bible, people had to read it right, to interpret its words correctly.

Out of the reading of the Bible grew an interest in the interpretation of texts and an appreciation that the written word, in whatever language, attains meaning only by interpretation. To determine the content of a text, readers have to know what the author intended the text to say. Learning about the intentions, background, and culture of the author could help readers understand the text.

These ideas on the interpretation of texts, on hermeneutics, were put forth in the early nineteenth century. We cannot begin reading without preconceptions about the text we are reading, what kind of text it is, what its general topic is, etc. In order to understand something, we already need a preliminary understanding of it. In order to ask a question, we already need a preliminary idea about the answer. Without such preconceptions, we cannot read, however open the book may be. These preconceptions will interact with the information we gain while reading, and they will influence our interpretation of the text.

We are telling this story about the book in order to draw attention to the analogy between the book and a computer system. Computer systems invite us to share the presuppositions of the French encyclopedists that information stored in an external device without indication of its author or context can be useful for whomever it may concern.

There are, however, limits to what can be made explicit. Insurance companies excel at using small print in making things explicit. But there will always be loopholes and unexpected cases not covered by the rules. Indeed, the basic idea of a bureaucracy is to operate only on the basis of explicit procedures. But for every rule that is added to a system of taxation, for example, the existence of exceptions and the need for further regulation will eventually be made clear.

In a famous dialogue by Lewis Carroll, Achilles is pushed by the tortoise into an infinite regress trying to make explicit the rules of reasoning in a simple inference. Let us write down the rule we are using here, says the tortoise. Achilles agrees. But then, asks the tortoise, what rule did we rely on in using that rule? Achilles writes again. And so on.

Carroll's dialogue argues that our use of rules always presupposes a practical understanding that cannot be made explicit. But notice that this should not discourage us in our attempts to design reasoning machines. As long as those machines follow the rules of reason, we need not worry that they lack the background understanding we demand of anyone following a rule. We can argue that this lack makes the machines regular rather than rule following, but that is all right as long as they make no logical mistakes. After all, they are only machines. The practical understanding that acts as background to our rules is relied upon in defending the rules, and we would not expect a machine to really be able to defend its rules.

Turning knowledge into information and information into data is a difficult task as soon as we aim beyond anything but the most formalized and routinized type of knowledge. Sometimes it is even quite impossible.

We have no idea how we do a lot of the things that we know how to do. Among those are the very fast feats of perception, recognition, attention, information retrieval, and motor control. We know how to see and smell, how to recognize a friend's face, how to concentrate on a mark on the wall or search memory for an old experience. And we certainly know how to walk and talk and some of us even how to sit down and shut up. But if someone asks us what it is we do when we do these things, we are at a loss. These are definitely tacit competencies. If there are rules involved, we have no idea what they might be.

This does not mean that cognitive psychologists are at a loss in theorizing about these mental processes. But it takes such a theory rather than an anthropologist asking questions if we want to design a computer system with this kind of knowledge. To the extent that playing expert chess, tasting wine, finding oil, or diagnosing pneumonia involves this kind of knowledge, our work as systems developers takes on new dimensions. Turning this kind of knowledge into information is not something we do in a four-month systems development project.

The fact that bureaucracies generate only more loopholes as they try to cover the loopholes in their rule systems and the fact that much of what we know is tacit and difficult or even impossible to make explicit should not discourage us as systems developers. Defining rules, identifying loopholes, defining new rules, and so on is a process that mirrors that of a developing science, and it is a good way to learn about reality in systems development.

But we have to realize that information systems will be used to support a practice in specific situations. When designing systems, we have to be aware of the way in which the system will change an existing practice, and we will have to be concerned with the difficulties involved in having the new system find its place in a new situated practice.

Platonic and Aristotelian Concepts

So far, we have looked at practical problems we face when wanting to turn knowledge into information. We can deepen our understanding of these problems by looking more carefully at knowledge itself. What is knowledge?

We can distinguish between *theoretical* knowledge and *practical* knowledge, between knowing that something is thus and so and knowing how to do something. We know the qualities and facilities of C as a programming language and we know how to program. When we claim to know that something is the case, we should be ready to defend our position, to argue for it, to answer questions like "How do you know?" "How can you be sure?" "What is your evidence?" When we claim to know how something is done, it is not the kind of knowledge we have evidence for, but we should be prepared to show that we can do it.

To know how to ride a bicycle, we must be able to identify bicycles, pedals, handlebars, and the like, but we don't have to know much more than that about bicycles. A small child or even a monkey can ride a bicycle. To identify a bicycle, we need the concept of bicycle, but our competence as bicycle riders does not demand a particularly rich concept. We are not expected to be able to define this concept or be able to argue for or against calling something a bicycle. In contrast, when we claim to know what a bicycle *is*, such abilities will be expected of us. All knowledge that something is the case demands the explicit and competent handling of concepts.

A common way of thinking of concepts is as packages of information. A concept is then some sort of object with a name and an internal structure, containing or implying rules of application, definition, and essential information about the things that fall under the concept. From Plato's theory of concepts as perfect instances situated in the world of ideas to current notions about objects in object-oriented programming, this is the dominant idea about what concepts are.

Bicycles differ in style and usability, they get wrecked and they rust. Models come and go. Our ability to identify them all as bicycles, and our ability to say what a bicycle is, depends on our access to the concept of bicycle, be it an idea in Plato's sense, that is, a perfect, unchanging bicycle in the realm of ideas, or be it an object in our mind or in a database.

To turn knowledge into information and information into data, we must model the concepts constituting the knowledge. This task is simplified when we realize that concepts differ in complexity. Most concepts, as Aristotle argued, are composed of other concepts down to a set of primitive, elementary concepts. Such Aristotelian concepts can be defined by enumerating their primitive elements. This sounds nice, but it is more easily said than done. Try asking yourself for definitions of such simple everyday concepts as bicycle, invoice, dog, or customer. Or what about concepts like computer, information, or knowledge?

When you try your luck on the concept of bicycle, you realize that one way to do it is first to divide the object into parts like wheel, handlebar, etc., and then to define those concepts. The first task is not unproblematic: How detailed should you be, where do you draw the lines of division? But the second task is similar to the original one: What is a wheel?

Why is it so difficult to give definitions for even the most mundane concepts? The obvious answer is that concepts are not primarily packages of information defined by structured complexes of more elementary concepts. But what else can they be? One interesting answer is that they are primarily defined by our practices and that we communicate about them primarily by prototypes or illustrative examples. In a way we are back with Plato, when we give this answer. I can tell if something is a bicycle or not by comparing it to a prototypical instance that I have access to. I don't have to consult a definition, a list of properties defining a bicycle, to identify something as a bicycle.

Aristotelian and Platonic concepts offer two different ways of making our knowledge and experience explicit. Using Aristotelian concepts, we formalize our knowledge by providing rules and criteria for determining when concepts apply. Using Platonic concepts, we provide typical examples that can then be compared to specific phenomena to assess similarities and differences. Aristotelian definitions lead to clear distinctions: The question whether a given phenomenon can be categorized as an instance of a concept is decidable. If and only if the phenomenon has the defining properties does it belong to the category. There is nothing to discuss. Platonic definitions lead to more fuzzy distinctions. We cannot always decide whether a given phenomenon is an instance of a specific concept. Attempts at making distinctions will often lead to discussions of similarities and differences between phenomena and prototypical examples and the personal background of the observer will influence decisions on how to categorize phenomena.

Many of our concepts seem to be of a Platonic character. This is particularly true of the concepts we acquire when learning how to do something, when learning a practice. On top of these prototypes we can develop Aristotelian versions, like adding verbal comments to images, whenever we begin to transform our practical knowledge into theoretical knowledge. The more we engage in talking and writing about things rather than doing things, the more adept we will be at Aristotelian conceptualizations.

But our expertise at doing things will not necessarily benefit from such transformations. Chess players, musicians, painters, diagnosticians, carpenters, tennis pros, and consultants all seem to excel at their arts without the use of Aristotelian concepts. When we want to transfer part of their expertise to computers, we cannot simply program the computer to imitate them.

Programming languages and programming techniques are very Aristotelian in character. In systems development, we shall have to transform Platonic concepts into Aristotelian ones, ending up with an expertise in doing something based on knowing something rather than the other way around. Our success in such an undertaking depends on the extent to which knowledge can be turned into information, implicit competence can be made explicit, prototypic images can be turned into words, and information can be turned into data.

One of the challenges of systems developers is to understand and respect the Platonic nature of human knowledge and communication, and to understand the computer not only as a machine for processing data based on Aristotelian concepts but at the same time as a tool to support human beings in using and communicating Platonic concepts.

The Romantic Challenge

The systems developer has the computer in one hand constituting a possible solution and in the other hand the concept of information posing the problem. The computer is the product of mechanistic thinking and a traditional computer science education is an education in such thinking. Doing systems development, it is at first natural to extend the idea of mechanism from the computer itself to its use, to information and its use in organizations. The idea of a bureaucracy, of an explicitly formalized organization working only with explicit information and Aristotelian concepts is such an extension of mechanistic thinking. The problem is only that there are no such organizations, and those organizations aspiring to be so are not very effective in dealing with a complex and changing environment.

As systems developers, we have to accept the fact that mechanistic thinking, so powerful in producing and characterizing the machine, may hamper us when we are trying to put it to good use. We have to see not only the mechanistic aspects of reality – reason, order, stability, and control – but also the reality and importance of power struggles, changing practices, and impending chaos. In short, we have to accept the challenge of romanticism, of taking more seriously the actual nature of organizations and their information use.

The mechanistic ideas of representation, formalization, program, order, and control add up to a coherent view of the world as a mechanism, and our thinking about it as mechanistic. But in spite of being powerful, or perhaps just because of its power, this world view soon becomes a straitjacket which people will try to break out of. In a mechanistic world, in a world of machines, there is not much room for God, the soul, and the living, or for creativity and artistic expression.

The mechanistic world view stood at its height at the time of the French Revolution. The intellectual leaders of that revolution were all mechanists. They were trying to bring order and reason, democracy and liberty into the world, by founding a society based on mechanistic, rational ways of thinking.

The revolution was hailed by optimistic intellectuals all over Europe as the birth of a new era. But the revolution soon went sour, the guillotine becoming its most prominent mechanism. Out of this turmoil came Napoleon Bonaparte, our first modern hero giving inspiration to new ways of thinking about the world, society, and man, to a romantic view of the world.

This romantic reaction to the mechanistic world view made much of the difference between organisms and machines, wanting to defend nature and everything natural against machines and everything artificial. Unconscious desire and uncontrolled emotional expression were at least as important as conscious reason and programmed mechanical reasoning.

The romantic philosophers were artists stressing the importance of individual artistic expression, of genius and creativity. Poetry rather than logic was their favored means of expression. To them, the universe was like a poem or a work of art, not a machine. The romantic philosophers were not interested in taking the universe apart like a machine, in analyzing it into its smallest atoms. No, they wanted to contemplate, understand, interpret, feel, and see through the world to its hidden meaning, as they might do with a poem or a painting.

When the romantics generalized this idea of expression, it turned into a metaphor with wide application. Everything – people, social institutions, technology, the universe – is an expression of something else, something deeper, something hidden behind the well-known surface. If the mechanistic philosophers of the seventeenth century tended to think horizontally, mapping the causal sequences of controllable machines, the romantics developed a vertical way of looking at things, always inquiring into the deep, uncontrollable forces behind everything.

The theories of Karl Marx (1818–1883) and Sigmund Freud (1856–1939) are obvious examples of vertical thinking. Freud made much of the romantic idea that human beings hide an unconscious animal behind their conscious, civilized surface. This romantic idea of "a beast within" inspired much nineteenth-century literature.

Listen, for example, to Robert Louis Stevenson in *The Strange Case of Dr. Jekyll and Mr. Hyde.* "I sat in the sun on a bench; the animal within me licking the chops of memory; the spiritual side a little drowsed, promising subsequent penitence, but not yet moved to begin. After all, I reflected, I was like my neighbors; and then I smiled, comparing myself with other men, comparing my active goodwill with the lazy cruelty of their neglect. And at the very moment of that vainglorious thought, a qualm came over me, a horrid nausea and the most deadly shuddering. These passed away, and left me faint; and then as in its turn the faintness subsided, I began to be aware of a change in the temper of my thoughts, a greater boldness, a contempt of danger, a solution of the bonds of obligation. I looked down; my clothes hung formlessly on my shrunken limbs; the hand that lay on my knee was corded and hairy. I was once more Edward Hyde."

Similarly, Marx used a two-story model of society: an economic base supporting a political and ideological superstructure. A study of man and society that stays at the conscious, political, and ideological surface will never reach a true understanding. Only by seeing through that surface into the unconscious, economic depths, will we be able to reveal the real forces shaping individuals and changing society.

Vertical thinking lends itself readily to that attitude of suspicion so forcefully practiced by the late romantic German philosopher Friedrich Nietzsche (1844–1900). If Nietzsche were to come across one of the computer-based information systems of today, his first questions would concern whose power interests such a system really serves.

The late eighteenth century brought new ideas to science and new machines. Mechanics made room for thermodynamics, the clock for the steam engine. The mechanists thought of change as the motions of a machine. Newton spoke of these motions as being caused by forces, but gravitation and inertia were humble creatures compared to the forces of electricity and heat. Not that the romantics needed these new ideas or machines to inspire them. Their political world was rich enough in revolutions and examples of uncontrollable forces.

The mechanistic dream of stability and order was replaced by a romantic fascination for disruptive change and chaos. The romantics saw change as the expression of hidden, bottled up forces, as eruptions resulting from the unleashing of mounting tension. Behind the calmest of surfaces, there may lurk a monstrous force ready to spring upon us. If the mechanists wanted to see only the surface, the romantics preferred to daringly court the monster.

Explaining change, the romantics used Plato's old idea of dialectics. For Plato, this was the method by which true knowledge was attained. We don't need more information, but we need to work more with the information we already have, and dialectics is the method by which this can be done. We see the method practiced by Socrates in Plato's dialogues, where a view on, say, justice, will be put forth only to be contradicted, and the result of all the arguing back and forth will be a more complex understanding of the original view.

The romantics took this method and turned it into a general theory of change as the dialectic expression of underlying contradictions. This is a powerful idea for understanding social change. It is obviously applicable in areas like science, business, or the practice of law, where fundamental conflicts, contradictory interests, and opinions define the practice.

The romantics wanted to think of everything in the universe as the expression of a single fundamental force, be it love, libido, or a will to power. Likewise, they preferred to explain a given process of change in terms of one fundamental contradiction. The latter idea will prove to be quite useful in our attempts to understand organizational changes.

Descartes distinguished between the way the world is and the way we perceive it. The mind is not just a passive receiver of impressions from the world. It shapes these impressions and imposes upon them its own mark. The romantics continued this discussion about our knowledge of the external world, distinguishing between the world in itself and the world of phenomena that we experience. Once we realize that smell, sound, and color are mere appearances, what is there to stop us from thinking likewise about shape, weight, space, and time? On what grounds can we say that we know anything at all about what the world itself is really like? Aren't we forever locked up in our own world of phenomena? Can we even tell whether there is a world in itself or not?

We are not going to pursue this line of reasoning here. But once we see that we, rather than the world, are responsible for how the world seems to us, the question of why the world is the way it is becomes a question about us rather than about the world. The early romantic philosophers followed the mechanists in trying to find an answer in our human nature, attributing the properties of phenomena to our sensory apparatus and to our mental faculties.

Later, the romantics began to argue that the concepts we use to order our impressions are cultural in origin. The world that we human beings experience, the world of phenomena, is constructed by concepts that vary from culture to culture and from time to time. Our ideas, rather than being representations of an external real world measured by their similarity to that world, are constructions measured by their internal coherence. Scientific theories are themselves constructions of worlds, not representations of a determinate real world out there.

The question of whether theories are representations of worlds or constructions of worlds becomes an important issue in systems development. Shall we look at a computer-based information system as a representation of the work processes in an organization, or rather as an attempt to construct an organization of such work processes?

A Dialectic Synthesis

The difference between the two world views – between mechanism and romanticism, between a world of light and order, computability, and representation and a world of darkness and revolutions, expressive forces, and construction – is like the difference between day and night. Engineers, of course, tend to be creatures of the day, while artists live only at night, or so they say. But even if as systems developers we have a strong leaning toward a mechanistic world view, we should not dismiss as nothing but poetic excesses a more romantic way of viewing the world. The romantic philosophers did not react so strongly to the fundamental ideas of mechanism as to the contradictions they perceived between these ideas and their implementations. Much disagreement between contemporary proponents of these two world views is rooted in a confusion of ideas with implementation.

The aim of science to attain a mechanistic understanding of the universe can of course be accused of demystifying the universe. If that was all the romantics had to say, we could perhaps take their accusations lightly, preferring reason to mysteries as we do.

But when our scientific understanding is applied in technological systems, making human beings subservient to unreasonable, autonomous machines, the romantic criticism of mechanistic ideas becomes more pertinent. Nothing will make us doubt the value of a democratic constitution, of gaining liberty by using rules to control the actions of the state. To that extent we are all mechanists. But when our rule systems profligate and our liberty is lost in a web of bureaucracies, we will listen more carefully to what the romantics have to say.

The implementation of information systems on computers raises many of these issues, and the discussion has followed the schema of raising romantic objections to a mechanistic world view. People have argued against the idea of computer-based systems as automatons, notably exemplified by early conceptions of expert systems, that computers should be kept in their place as tools. People have also argued against the idea that such computer-based systems can be used to control the work processes in organizations, that organizations cannot successfully be treated so mechanically. And finally, people have argued against the idea that the end of a systems development project is to produce a system that gives an adequate representation of vital information flows in the organization, that the aim of such a project should be to organize a process of change involving all the members of the organization.

Of all the romantic ideas that we find in recent discussions of computer usage, the major role is perhaps played by what Nietzsche called perspectivism. When the romantics gave up the attempt to formulate the final truth about the world and began to see concepts and conceptual schemes as constructive creations of worlds, they introduced the idea of different perspectives. Rather than looking for the true perspective, the romantics wanted to use different perspectives to open up different dimensions of reality.

Take the computer as an example. What is a computer? How should we view it? Is it a computing machine? A control system? An expert? A tool? A medium for communication? A system for documentation? With a romantic approach, we are able to use all of these perspectives to aid us in our understanding and use of the computer. With the addition of new perspectives, we see new things in the computer, find new uses. It even makes good sense to say with the romantics that the computer is what we conceive it to be. As designers of computer technology and information systems, we must take into account the perspectives of the users of the technology. Those perspectives play a vital role in the construction and reconstruction of the technology and its use.

Nietzsche used his perspectivism to argue that theoretical positions in science, favorite metaphors, tell more about the background and interests of their proponents than they do about the world. A perspective such as "the computer is a tool" is put forth, objected to, and defended with such vigor not because people worry so deeply about truth but because they have interests to defend.

In our attempts to answer the question "What is information?" we have learned that there are no simple answers. Making knowledge explicit and transforming information into data are at the very heart of systems development. But as we go to work defining Aristotelian concepts, we cannot afford to forget to pay homage to Plato. It is not just a matter of choosing between bureaucratic and organic approaches. In designing organizations, information systems, and computer systems, the issue is not whether to formalize or not. The challenge is to find an appropriate degree of formalization.

From a mechanistic standpoint, formalization and Aristotelian concepts are necessary in order to arrive at a truthful understanding of the world, and they are needed in order to avoid misunderstandings in communicating about the world. To a romantic, this is a simple and naive view that does not take into account the dynamic and chaotic nature of the world and fails to deal constructively with different perspectives and conflicting interests.

Real organizations and their information systems are mixtures of different approaches to organizational design. Some are more bureaucratic and formalized in nature, others more organic and less formalized. But none of them operates according to a single strategy. Even if we attempt to design an ideal information system based on a single strategy, the actual practice of using the system will mix the two strategies.

Systems developers have to be able to combine a mechanistic understanding of computing machines with a romantic appreciation of the complexity of human beings, social organizations, and information use. But how do we obtain such a combination? Can these two, in many respects diametrically opposed, world views be merged into one? Our answer is clearly yes. But our idea of how it should be done is not the one suggested by the above formulation: thinking mechanistically about computers and romantically about people.

We began, in chapter 1, with a mechanistic understanding of computing machines, which was later embedded in a world view and related to our understanding of organizations. As we then moved on, in chapter 2, to discuss computer technology use, this mechanistic world view came up against the complexity and uncertainty of the social world, its organizations, and its use of information. Finding the mechanistic world view unsatisfactory in dealing with the use of computers, we began to introduce romantic ideas and eventually a whole romantic world view.

These two world views are clearly contradictory, but they can enrich one another only if they are allowed to encroach on each other's territory. It is when we confront a mechanistic view of organizations with a romantic view, or a romantic view of computing machines with mechanistic ideas that interesting things begin to happen.

If the two world views are simply placed next to one another, nothing of interest will happen. It is in the dialectics between them that we can hope to find the interesting synthesis of mechanistic and romantic ideas. It is when we introduce dynamics and change into the allegedly stable and well-ordered world of machines, or formalizations and order into the chaos of human affairs, that we are on our way toward an interesting synthesis of the two world views.

Such a synthesis will remain unstable. The dialectic struggle between its contradictory elements will continue. But it is on this pattern that we have tried to write this book: A mechanistic understanding of something will be presented, later to be confronted with a romantic challenge, the two eventually merging into a dialectic synthesis. We will not reject our fundamental allegiance to a mechanistic world view, nor will we try to avoid meeting the romantic challenges. This pattern will be repeated several times, each time adding, we hope, more material to a deeper understanding of the ideas involved in the practice of developing information systems, the practice of putting computer technology to good use.

In this process, we shall have to abandon the idea of the computer as a solution to a problem. To begin with, it is not clear what a computer is. Depending on our choice of perspective, on the metaphors we use to describe the computer, the solution suggested by the computer is very different. And in many situations it is difficult, or even impossible, to agree on what the problem is. Even if people seem to agree, it can be quite difficult to define the problem. And if a problem eventually is defined and agreed upon, we tend to change our interpretation of the problem as we go along.

Computers are not used to solve problems. Computers are used as a means to change organizations. Systems development projects are initiated because some actors perceive the present situation to be problematic, and they have hopes for a change involving the use of computers. Afterward problems are seldom solved. But the situation is different. At best, this new situation is more satisfactory than the original one, even if it too is problematic, raising new challenges.

3

Thinking

Hard Systems Thinking
Soft Systems Thinking
Dialectic Systems Thinking
It Makes a Difference

We have talked about different ways of conceiving of computers, and we have discussed the concept of information. But we have said nothing about processors or compilers. Nor have we discussed bits or information channels. This is not because we think an understanding of the basic technology and of information theory is unnecessary. Quite the contrary.

Systems developers need a thorough acquaintance with information technology in order to be able to put it to fruitful use. But we reject the simple idea that it is enough to learn what is worth knowing about computers and information and then base our practice on this knowledge only. Systems developers need frameworks for thinking within which they can apply their knowledge to the challenges they face in their practice.

We all acquire such frameworks in our education and practice. These frameworks are expressed in the theories, methods, and concepts we use, and they are reflected in our views on the problems of our profession. But to what extent are the frameworks we rely on actually confronted with our practice? To what extent are our fundamental ways of thinking open to discussion and subject to change?

When talking about our profession, we cannot avoid the word *system*. Systems development is the business of constructing computer systems for the use of human beings in receiving, processing, storing, and communicating information. The computer system and its immediate users constitute a wider information system supporting or controlling the activity of the larger organization, which we naturally come to think of as yet another system.

If this is how we understand systems development, we have implicitly accepted a systems approach to computers, information, and organizations. There are, however, different kinds of systems approaches. In this chapter we shall present three such varieties. The hard systems approach emphasizes clear, exact, and true representations of the world. The soft systems approach pursues the idea that there are always several, equally plausible perspectives of the world. Finally, the dialectical systems approach is based on the idea that to understand, explain, and make possible change we must think in terms of contradictions.

We shall see that the three approaches differ radically in the way they support us in understanding, designing, and changing the use of computers and information in organizations.

Hard Systems Thinking

A hard system is a hierarchically organized set of elements. Elements on a given level are composed of elements belonging to the level below. The properties of the elements of a given level are reducible to – that is, definable in terms of – the properties of the elements on the level below. Everything that can be said about the system can, in principle, be expressed in terms of the elements and properties at the lowest level of the system. As a consequence, the hard systems approach puts heavy emphasis on the internal structure of systems.

The periodic system in chemistry is an outstanding example of the power of such structural decomposition. More generally, the hard systems approach has pervaded modern natural science with its analysis of material bodies into molecules, atoms, elementary particles, and so on. It has encouraged the idea that physics, being a representation of the elements and properties at the lowest level of the system of nature, is a general and all encompassing science of the world: Everything there is to say about the world can be said in the language of physics.

For hard systems thinkers, a system is typically a functional system, a machine with a determinate function. Our method of understanding relies on taking the mechanism apart to see how it works. The functional analysis of a machine divides it up, level for level, according to the functions of its parts, disregarding properties that are functionally irrelevant, treating as equal properties that are materially different but functionally equivalent.

Functional analysis is used to deal with the complexity of the world. A system is defined and demarcated from its environment by its function, and the system itself is analyzed in terms of the functional roles played by its elements and their properties. The bewildering complexity of the world is handled by dividing it up into systems, each system being further analyzed from the top down to its lowest functional elements.

By concentrating on the question "What does this thing do?" we abstract from all properties that are functionally irrelevant, thus reducing the complexity. From a practical point of view, hard systems thinking supports us in understanding the world by providing us with this explicit criterion of functional relevance.

Functional analysis is both a method for understanding and a method for design. Designing a functional system, we ideally proceed from the top down, asking first what the system as a whole is supposed to do and then working ourselves down, level by level, to ever more simple functional subsystems until we reach elements that we know how to construct, or already have at hand.

Functional analysis is universally employed in methods for systems development and programming. Data flow diagrams, stepwise refinement, and top-down development are all based on this powerful idea. Functional analysis – and hard systems thinking – is the classical approach to programming and systems development.

Think of a simple management information system for the distribution of books from a publisher. Applying hard systems thinking, the function of the system is to maintain information needed to distribute books to customers. This includes functions like order entry, invoicing, distribution management, inventory control, administration of customer information, and accounting.

On the most abstract level we can identify those events in the system's environment that the system must respond to, and we can describe the data flows connecting the system to these events. Typical events include customer orders book, customer sends payment, customer wants refund check, warehouse receives book returns, and warehouse makes shipment.

On this level we can describe the function of the system simply as the administration of book distribution. Having identified this overall, abstract function and its relation to its environment, we can then decompose the system into a number of more concrete functions. These include functions such as process orders, manage customers, produce accounting records, and interact with warehouse. All the functions on this more concrete level are then related by internal data flows under the condition that their total effect equals the already agreed upon data flows on the higher level specifying the relation between the system and its environment.

We continue this process of refining our understanding until we reach the most concrete level of functions needed to specify the behavior of the system. This lowest level of the system includes functions such as find out if book is in stock, enter order, check credit authorization, and process salesperson order. Each of these functions are then specified as simple procedures, and, in addition, we include in the data dictionary detailed descriptions of all the data involved.

This example, borrowed from Edward Yourdon's *Modern Structured Analysis*, illustrates hard systems thinking as it is practiced, using context diagrams, event lists, data flow diagrams, process specifications, and data dictionaries.

The exact sequence in which the different levels of the system are described varies. A radical strategy would invite us to iterate between different levels. A conservative strategy would force us into a more strict top-down approach. Whatever strategy we apply in specific situations, we practice hard systems thinking.

In the hard systems approach we proceed on the assumption that reality is itself an ordered, stable system. Our goal is to find the true representation of the world and from then on our efforts concern the representation rather than the world itself.

The role of the systems developer is to map what is there and to ensure the truthfulness, consistency, completeness, and implementability of the system. We start out from an existing system or from an idea about a possible system and we end up with a computer realization of this system corresponding to a specification of it.

For systems developers, this means that the world plays its role as a source for creating specifications, but it is not a concern in itself. A functional analysis of the system results in a specification of requirements, which from then on provides the quality measure for the design efforts. When specifications are met, the work is done.

The system is out there, in the world. Its boundary and function are objective, for us to discover and analyze. Our job is to represent the system truthfully in our minds by constructing a corresponding symbolic system. We can do this graphically by drawing a tree-like structure mapping the structure of the system. Or we can assign to every element a name, to every property a predicate, constructing sentences isomorphically matching the system.

This idea of sentences as representations of facts, language as a picture of the world, played an important role in the development of modern logic. Much of the enthusiasm created by formalized languages, particularly first order predicate calculus, was based on the idea that this language could be used to give an isomorphic representation of the world. This representation would be objective in the sense of admitting only one interpretation given by the unique assignment of names to things and predicates to properties.

"The limits of my language mean the limits of my world," was how the young Ludwig Wittgenstein formulated the idea that language is an isomorphic picture of the world. He was not thinking of just any language but of the ideal language that he believed first order predicate calculus to be.

Whatever we say of this dream of constructing a language formally equivalent to the structure of the world, that dream still plays a role in the development of computer systems. In forming Aristotelian concepts, in evaluating specific programming language constructs, and in reviewing program specifications, we constantly rely on this idea of structural correspondence between our representation and the real system. Only rarely do we seriously question the underlying idea that there can be a perfect, truthful in the sense of isomorphic, representation of a real system.

When systems developers conceive of a work process as represented by a data flow diagram, the powerful dream of a language in which everything can be said easily turns into a barren vision of what the world is like. Practicing hard systems thinking, we have to be constantly aware of this danger of getting trapped by our current means of representation.

The idea that there must be a true representation of the world is easily confused with the belief that our representation is the true one. Similarly, an interest in stable and ordered systems makes us unwilling to accept change. Once we have constructed a representation of a system, we prefer to treat changes in that system as aberrations from truth rather than as demands for further constructive representation.

There is a joke about military instructors advising their students to assume that "when the map and the terrain disagree, the error is in the terrain." This has become something of a cliché warning us against getting caught in our representation of the world. There is a nice cartoon elaborating on this idea. Three men are standing on a mountaintop in an area crowded with similar mountains. They are bent over a map, looking worried. Suddenly one of them raises his head, pointing to a mountain in the background, saying with relief: "Now I know, we must be over there!"

Soft Systems Thinking

We find it both natural and practical to talk about computer systems, information systems, management systems, documentation systems, and so on. But in using the word system in this way, we implicitly accept one of the basic assumptions of the hard systems approach: that the system is out there. Out there we have information systems, and in here we have descriptions of those systems.

What if the systems are all in here, in our minds? Perhaps, there are no systems out there, in the world. Perhaps, we are just conceiving an aspect of the world as a system. The multitude and variety of phenomena we are dealing with in the world are so great that our conceptions can be nothing but poor and simple substitutes for the real thing.

When we say "this is a system," we really mean to say "let us conceive of this part of the world as a system." When we say "this is a computer and that is a table," we really mean to say "let us conceive of this as a computer and that as a table." If this is what we mean, why don't we just say it? One possible answer is that it is more practical to use the shorter and less ornate expression. Another answer is that our daily language really reflects the way we think. The reason that we use the word system as we do is that the hard systems approach dominates our profession.

The soft systems thinker makes a major point of the observation that our world is shaped by our experience of it. We see different things, have different perspectives, structure the world differently, depending on interests, background, education, and culture. The world we perceive is the world we live in. Our world will change if our perception of it changes, if we develop a new way of looking at it. The world we experience is more interesting than the real world. To us it is the real world.

Systems thinking is a way we have to deal with a world that is diverse and constantly changing. The world is constructed by us in our perception of it. The order we perceive is of our making. When we see systems in the world, they are the result of our attempts to organize our experiences, beliefs, and visions. We see what we believe, rather than the other way around. A system is based on assumptions about the world. Different assumptions give different systems. As a consequence, there are always several perspectives, resulting in different systems, on the same concrete situation.

To hard systems thinkers, systems are out there, and we build them, change them, and improve them, by engineering. We see them, and believe what we see. To soft systems thinkers, systems are in our minds, they are perspectives that we change and improve by being confronted with other perspectives, by getting around in the world and experiencing new things, by learning.

But learning is difficult, and it is especially difficult when it means changing our very conception of the world, our basic assumptions about what is important and what is not. Confronted with people with radically different perspectives on the world, our natural reaction is one of noncomprehension. We simply cannot understand what they say. What to us is terrorism, they see as a fight for freedom. What we know is simply a communication problem, they say is a power struggle.

A soft systems approach must develop a methodology for helping us understand perspectives that differ from our own. The idea of interpretation is as important to soft systems thinking as the idea of truthful representation is to the hard systems approach.

A soft system is a totality with emergent properties. Moving up through the levels of the system, new properties emerge. A soft system is a holistic perspective that cannot be defined in terms of its lowest level elements and their properties. An operating system, conceived as a totality, has properties such as the deadlock property, which cannot be understood on any of the lower levels of the system.

When deadlock is viewed as an important property of an operating system, we are not simply describing a system as it is. We are expressing a perspective based on the assumption that an operating system is meant to allocate resources efficiently and with fairness. As a consequence, deadlock should be avoided. Supporting this perspective, we find other assumptions about the users of the system, about the kinds of services they are asking for, and about the kinds of restrictions they will accept as users of shared resources.

What a hard systems thinker views as a single operating system will become many systems when viewed by a soft systems thinker. These different systems may lead to different evaluations of the efficiency and fairness of the operating system. And they may provide us with different strategies for improvement, modification, and change.

The method of the soft systems approach is interpretation. We are encouraged, by this method, to consider different perspectives; the claim is that to learn about the world we have to understand, express, and debate a variety of radically different perspectives.

To improve the way work is organized in a systems development group, we cannot rely only on the abstract system that is expressed in the standard project model used by the group. For one thing, we should compare this system to the beliefs and attitudes of the project members and learn from the differences between the ideal world of the project model and the experiences and ideas in the projects. But more importantly, we should formulate alternative systems, expressing the perspectives of the involved actors on issues such as project organization, communication and cooperation, programming, testing, contractual arrangements, and project management. We should develop project models based on such systems, and compare them with the actors' views on present and future practices.

The managing of a systems development project can in this way benefit from soft systems thinking. This approach offers a way for the involved actors to express and debate perspectives on their situation in their attempts to find possible improvements and changes.

Proponents of the soft systems approach often complain that we don't think systematically about our beliefs and visions and that we don't confront them with our perception of present practices. In designing more effective organizations and information systems, we can learn a lot by comparing our beliefs, formulated as systems, with our actual practices as we perceive and experience them.

The users say and think that the information provided by the computer system supports them in making decisions. But a systematic comparison between explicit system models and concrete experiences might reveal that even if much effort is spent on updating the system, the information provided is so unreliable that the system is seldom used in making decisions. Most systems developers say that quality assurance is important. But looking closer at their actual practice, we will often find that it plays a marginal role in their work.

Another complaint is that we tend to be blind to differences in perspectives. We go around taking for granted that people view the world as we do. Sure enough, we know that there are all kinds of crazy strangers, but it generally comes as a surprise when people at our own workplace turn out to be pacifists, Buddhists, vegetarians, or even Catholics.

The soft systems approach reminds us that we are different and that different perspectives makes a difference. It offers a strategy for expressing such different perspectives and for engaging people in a debate with the purpose of reaching some sort of agreement that can lead to the formulation of requirements for action. Systems are used to make perspectives explicit, and through debate our experiences become clearer by being confronted with such systems.

To illustrate, let us return briefly to our example of the simple management information system for the distribution of books. Applying soft systems thinking, or, more specifically, Peter Checkland's *Soft Systems Methodology*, we would engage ourselves in the present situation in the publishing company.

Together with the involved actors, warehouse staff, marketing people, managers, administrators, and computer personnel, we would make explicit their different views on the company, their different roles in daily routines, and the various kinds of human activities involved in the distribution of books. We would try to get a feel for the situation without enforcing general concepts and personal preconceptions. We would express our impressions and findings in a number of rich pictures, each containing a personal interpretation of the situation.

The situation is then evaluated and a number of possibly relevant activity systems are identified. We might decide to look more systematically at ways to give better service to existing customers, to get new customers, to process orders effectively and efficiently, to provide information about available books, to manage the stock of books, etc.

Now we leave the unstructured real world and enter the world of systems thinking. Some of the identified activity systems are analyzed one at a time. First, each system is defined by a root definition and then it is described in more detail by a conceptual model.

The root definition defines the main characteristics of the system: the customers, the actors, the transformation, the underlying assumptions (Weltanschauung), the owner, and the environment. The conceptual model represents the system as a set of logically related human activities together constituting the essential transformation as identified by the root definition.

It is important to notice that the root definition is a specification of an activity system, not a solution to a problem. It is a common misunderstanding, we have found, to interpret Checkland's methodology as a general method for problem solving. The problematic situation is interpreted as a problem and the root definition as a solution to that problem; in reality, the root definition is an attempt to make a fresh start, abstracting from the actual situation and using rational thinking to develop a systematic idea of how the business in question should be organized.

Leaving now the structured, rational world of systems thinking, each abstract system is brought back to the situation and confronted with the actors' experiences, beliefs, and conceptions. Each system is an abstract phenomenon expressing a possible way to perform some of the activities needed to distribute books in the publishing company. The different systems might be overlapping and even conflicting, but each is debated among the involved actors. Does this model correspond to how we do things today? Are we satisfied with this way of doing things? What kind of improvements can we suggest? How do we evaluate the effect of such changes?

Some systems will prove more useful than others. Some will lead to many proposals for change. Others might prove worthless in this respect. The process is highly iterative and participative. The manifest outcome is a list of proposed changes for improving the way the publishing house manages the distribution of books. Some proposals might include the use of computers. Others might focus on organizational changes.

In the soft systems approach, design is seen as learning. We start with an unstructured situation and end up with requirements for change. These requirements are the result of a learning process where perspectives are formulated and organized into systems. These perspectives are communicated, compared, and negotiated with the purpose of reaching a common understanding and a shared interest in change.

Soft systems thinking provides us with a rich and realistic approach to learning and change, but it is difficult to plan and manage. There are no guidelines for how many perspectives will suffice, and the only way we can learn about the relevance of a perspective is by making the effort to formulate it explicitly as a system and then to debate it. Moreover, this approach does not in any constructive way support implementation. The result is a list of negotiated requirements, but how are we to go from the present situation to an improved situation satisfying these requirements?

In short we can say that soft systems thinking provides us with a rich and realistic approach to learning and change. But this is achieved at the expense of a complex and uncertain process, requiring substantial experience and professionalism by its practitioners.

Dialectic Systems Thinking

In soft systems thinking, multiple perspectives are seen as a basis for fruitful discussions and deeper understanding. Alternatively, different perspectives can be seen as expressions of, and the means in, an irreconcilable conflict and power struggle. Perspectives then become expressions of different interests and inherent contradictions – for example, between systems developers and users or between managers and employees.

The dialectical systems approach is based on the idea that the world is always changing, and that we cannot understand it unless we understand what change is and why it takes place. The claim of the dialectical approach is that we must think in terms of contradictions in order to understand, explain, and control change.

Some conflicts are founded in misunderstandings and ignorance, and they can be resolved through communication and discussion. But other conflicts cannot be resolved in this way. These conflicts are social expressions of underlying contradictions. Resolving these conflicts requires more substantial changes – not only changes in the actors' perception of the situation but, equally important, in hard matters like the distribution of resources, organizational structures, technical systems, and power relations.

It is fairly easy to see the role of contradictions in our thinking. We rely on contradictions to determine whether we should change our views or not. When a colleague in a systems development project contradicts us, we engage in discussion. When one assumption underlying the design of a program is incompatible with another, we change the design.

But in the dialectical systems approach, contradictions appear not only in our thinking but in the world itself. Reality is assumed to be a totality of related contradictions, its most dominant feature being change. In any given situation we face a network of related and dynamically changing contradictions. Some of these are more fundamental than others, and in each of them one of the two opposites is more dominant than the other. As the situation changes, other contradictions will become more important, and the opposites of each contradiction will be differently balanced.

In a bank engaged in developing its next generation of a computer-based transaction system, there might be contradictions between a well-functioning centralized computing department and the distributed nature of the new system, between the physical design of the branch offices of the bank and the way in which work is to be organized in relation to the new system, or between the requirements and expectations of the new system and the experience and competence of the systems developers.

At each stage of development, some contradictions will seem more important than others, and this will be an indication of possible interventions. Should the branch offices be redesigned, or should work be organized in a different way? Should the systems developers receive further training? Should additional systems developers with different skills be employed? Should management and users realize that their expectations and requirements cannot be met?

Returning for the last time to the publishing company, let us once more think about the simple management information system for the distribution of books. Applying dialectic systems thinking will not add a third kind of systems to our thinking. Instead, the dialectic approach will help us attend to contradictions in the company involving both hard and soft systems.

As in the soft systems approach, we begin by getting a rich feel for the situation in the publishing company. Analyzing and evaluating the situation, we identify interests, roles, structures, and processes, and we identify important contradictions related to the distribution of books.

We might focus on the contradiction between having a limited amount of capital invested in stock and the need for quick deliveries to customers. Or, we might find a contradiction between a need for standardized information about books, authors, and customers, and the different information needs of accounting, marketing, stock control, editorial work, and management.

In the technological area we might be concerned with the emerging conflict between users wanting a network-based architecture with workstations and the established traditional technical platform. We might also focus on the contradictions between the existing central organization of the computing staff and the need for new and different kinds of computer expertise and a more distributed organization of these.

Each of these contradictions is analyzed in detail. Major areas of conflict, the interests of the involved actors, and the dominating side of each contradiction are identified. Possible negotiations, compensations, or compromises that might help loosen the situation and support further development are considered. Prospects for alliances, different types of interventions, and suggestions for change are examined and evaluated.

These considerations are used to select a strategy. Actions will be performed and the situation will change, as will our conceptions and beliefs. Having started this way, we find ourselves engaged in a dialectic process of thinking, trying, rethinking, and trying again.

When we use a dialectical systems approach, we must pay particular attention to the relation between reality, as we understand it, and the different perspectives people have of reality. We are not primarily interested in these different perspectives, as we would be with a soft systems approach. The world, rather than people's perceptions of it, is our primary source for learning.

Instead of arranging debates between proponents of different perspectives, the dialectic systems thinker will use contradictions as real grounds for action. But in contrast to hard systems thinking, the dialectic approach takes seriously the different perspectives, viewing them as expressions of a world that is inherently dynamic and pervaded by contradictions.

In the dialectical systems approach, design is action. We can make contradictions explicit, and we can negotiate perspectives. But we cannot know the meaning or the importance of a perspective without testing it in practice. The basic idea is to understand and explain change by contradictions and to learn about possible changes through intervention and action.

It is easy to think that dialectical systems thinking is only applicable to the social world. Since there are no actors or conflicts within a computer system, it is only fair to question whether it makes sense to speak of contradictions within such a system.

Contradictions have been used, however, to describe not only the social world but the psychological world as well. We are thinking of Sigmund Freud's psychodynamic theory of the mind. According to Freud, our minds are made up of three subsystems: the superego, the ego, and the id. Each subsystem has a different view of itself, of the mind as a whole, and of the world.

The superego is the perspective of an internalized father figure, it is the conscience, the voice of God within the individual. The id is a roaming beast ("the animal within") constantly desiring to devour whatever comes in its way. The ego is the poor middleman, trying to remain sane while negotiating between the superego and the id, using much mental energy, often having to bind the two parties in an energy-consuming stalemate. The more energy a contradiction consumes, the more it will be said to dominate the system. The therapeutic task of the psychoanalyst is to release that energy by helping the ego to resolve or handle the contradictions between the id and the superego.

With this example as an analogy, we can reflect upon the contradictions between fairness and efficiency in an operating system, between general applicability and expressive support in a programming language, between search and update of a database, or between ease in using and ease in learning in the design of interfaces.

When working with computers, we are seldom faced with simple problems having unique solutions. We are typically faced with what are traditionally called trade-offs. From a dialectical point of view, these trade-offs are manifestations of contradictions inherently related to the use and development of computer systems.

Hard and soft systems thinking are both well-established, relatively clear, and well-developed ways of thinking about the world. They have been thoroughly debated and confronted with one another again and again in our intellectual history, going all the way back to the Greeks. In the last 200 years they have been poised against each other as alternative approaches to social science research.

The dialectic systems approach has no comparable position in our history. With its roots in the rather obscure philosophy of Georg Wilhelm Friedrich Hegel (1770–1831), dialectic thinking has been mainly used by Marxists in a way that has often made people wonder and shake their heads.

We shall not use this occasion to try to formulate a well-rounded and complete alternative to the hard and soft approaches to systems thinking. But there is in Hegel, and in dialectic thinkers such as Marx, Nietzsche, and Freud, a rather simple observation, and once you have shared it you can no longer rest content with choosing between a hard and a soft approach to systems thinking. The observation is that life is a struggle.

Both hard and soft systems thinking share a wonderfully naive view of the world as fundamentally harmonious. Either the world itself is well ordered and stable and it is our task to represent it truthfully and to organize our societies and lives on its model. Or else, there are many possible perspectives on the world, but they can all be joined in a rational dialogue, with a mutual interest in reaching the final synthesis of agreement, truth, and understanding.

This way of thinking of the world dominated Western thinking all the way up to the nineteenth century. But this is not the world we live in. Our world is not a harmonious, stable, well-ordered, rational, peaceful world with mutual interests, agreement, and understanding. Our world is bordering on chaos, suffused with conflicts and dark, irrational forces, a dangerous, frightening jungle of different interests and struggles for power. And there is an alternative conception of reality, developed in the nineteenth century, exemplified by Hegel's master-slave dialectic, Darwin's struggle for survival, Marx's struggle of the classes, Nietzsche's will to power, and Freud's neurotic individual.

It is this conception that we want to do justice to in what we call the dialectic systems approach. More than the others, this is our fundamental approach to systems thinking. The dialectic systems thinker accepts the ambition of the hard systems thinker to map the world, as well as the ambition of the soft systems thinker to have a rational debate between different constructions of the world. But it questions the underlying assumption of harmony, order, and common interests. By adding a dimension of conflicts, contradictions, interests, and struggle for power, the dialectic thinker will look on these ambitions with suspicion while still carrying on the show.

It Makes a Difference

The hard systems approach is hard in the sense that it relates systems to matter. It finds its scientific support in the natural sciences and in mathematics, and, using its method of *analysis*, it is effective in dealing with the complexity of a system that is out there as part of a stable and harmonic world. Systems thinking is a process in which we compare our representation with the system, working our way toward an ever more truthful representation by observing the system and by developing our means of representation.

As hard systems thinkers, we will think of organizations and applications of computers as systems with a function. We will take pains to define these systems and their function and then go on to decompose the system into subsystems having different functions contributing to the function of the whole system. We will think of organizations as machines, and our task will be to make such machines more efficient, better at fulfilling their function.

The soft systems approach can be viewed as an attempt to transfer the systems of the hard systems approach from the world to our conceptions of the world. It is soft in the sense that it relates systems to our mind. It turns to the humanities and the social sciences for supporting theories and ideas. When going from a hard to a soft systems approach, we seem to lose in elegance what we gain in richness, and we seem to lose in constructive power what we gain in realistic insight.

Soft systems thinking with its method of *interpretation* is not focused as directly on complexity. Uncertainty has become an important challenge. We are no longer confronted with a definite but complex system. Instead, we are confronted with an unstructured situation and with different perceptions of that situation.

If hard systems thinkers run the danger of getting caught in their means of representation, mistaking their constructions of reality for reality itself, then soft systems thinkers make a virtue of this danger by stressing the fruitfulness of comparing different constructions of reality with one another. By engaging proponents of different perspectives in a debate, the aim of soft systems thinking is to arrive at a richer, well-argued set of requirements that the proponents can agree upon as a useful starting point for change. By stressing the importance of discussion, of comparing different perspectives with one another rather than a single representation with the world, soft systems thinkers express their preference for the communication of ideas and mutual understanding.

As soft systems thinkers, we will be interested in learning about the organization and the application of computers by making explicit the views and beliefs of the actors within the organization. We will think of the actors within the organization as acting rationally on the beliefs they have about the organization and about the use of computers. By taking their different perspectives seriously, we will let them do our work for us, or so a hard systems thinker might describe it.

The dialectic systems approach is a dialectic synthesis of the other two. It takes both reality and perspectives seriously, both matter and mind, both mathematics and the humanities. But it places all this in the context of contradictions, interests, and struggle.

With a dialectical systems approach, we see the organization as very different from the stable and ordered system of hard systems thinking. Thinking dialectically, we stress contradictions and change, conflicts of interest, and struggle for power and influence. The different perspectives we find within the organization are taken as indications of such contradictions and conflicts. Our interest in such different perspectives is motivated by our interest in identifying contradictions and conflicts in order to use them as our basis for action.

We no longer believe in reaching consensus through communication. Or rather, we believe that such consensus will only be temporary unless the underlying contradictions are resolved by *action*, by changing the organization or the computer system itself.

A dialectical systems approach is both soft and hard. We use different views to identify real and relevant contradictions within the organization. We think that organizational actors tend to overlook fundamental contradictions, and we think of organizations as a place where too much energy is spent on defending positions, neutralizing conflicts, and negotiating interests. We see our task as one of utilizing and freeing some of that energy to design and develop useful computer systems.

We have outlined three different ways of thinking about computers and organizations, each implying a different role of the thinker. In the hard systems approach, the thinker is an outsider who gets information about the system by observing it or imagining it. The primary task is to represent these observations or imaginations in a well-structured manner. Metaphorically speaking, hard systems thinkers are photographers taking pictures of the systems out there.

In the soft systems approach, the thinkers are part of the situation with which they are concerned. Either they have been called there as consultants for the involved actors or they might be involved actors themselves. They learn about the situation by explicitly formulating possibly relevant perspectives, by debating these perspectives and comparing them to the situation. Their primary task is to facilitate learning and to engage the involved actors in this learning process with the purpose of creating a useful and acceptable platform for change. Metaphorically speaking, soft systems thinkers are teachers.

In the dialectical systems approach, the thinkers are outsiders and insiders at the same time. Dialectic thinkers are arrogant enough to try to combine the two roles. They get information by uncovering contradictions in the situation, but they are also involved actors. They are engaged in debates about interests and perspectives, and they negotiate possible actions. Their primary task is to understand the given conditions and options and to engage in a cooperative and political process of influencing and changing the situation. Metaphorically speaking, dialectical thinkers are change agents, political actors.

Going on and on about the different approaches like this, comparing them to one another, becomes rather boring. And without realistic examples, the approaches will remain unclear. Suppose you have been given an open invitation to deliver a proposal for how computer technology could be used in a public high school. Suppose further that you have the time to make a thorough analysis of the high school as an organization. Where would you go looking? And what would you see? Take a minute and think about this before you read on.

Would you spend most of your time with the school board, studying legislation and curricula covering education in public high schools? Would you study the organization, the use of classrooms, the division of labor between the teachers? Would you try to define the function of a public high school in relation to students, teachers, future employers, or society at large? Would you then look carefully at how the different activities in the school combine to fulfill that function? Would you come up with ideas for rationalization, for better organization of resources, better fulfillment of goals? If you would, you are a hard systems thinker.

Or, would you jump right into the classroom, engaging students and teachers in a discussion about the school? Would you organize discussion groups with students and teachers and members of the board, stage panel debates and happenings? Would you confront students and teachers with possible uses of computer technology, showing them possible systems and setting up study groups in the use of computers? If you would, you are more of a soft systems thinker.

Or, finally, would you take a hard look at what is *really* going on in schools? Would you try to show, for example, that the school is really an institution for disciplining both students and teachers? Would you show how all the major institutions of modern society – schools, the army, mental hospitals, churches, factories – share with prisons a fundamental communication structure for the control of its members?

In all these institutions, communication travels in one direction – from some sort of power center to the "inmates." In the classroom the teacher speaks. The students are requested not to talk with one another, and questions to the teacher are kept at a minimum, in spite of declarations to the contrary. There is never enough time to deal with such questions. Students are made to feel embarrassed whenever they take time from the scheduled work, so they learn to keep quiet.

Would you use this kind of analysis to reveal hidden agendas, to argue that when the school is teaching students geography, it is using the geography lesson as a pretext for teaching students their place in society? And would you go on to analyze the power relations between parents, teachers, students, the school board, the principal, and the community, and use these analyses to intervene in the process of education, taking sides, making allies, bringing deep contradictions and conflicts into the open?

Well, would you? Tough luck if you would, for you would be a dialectical systems thinker and your work would be in constant turmoil. God knows when you would be able to deliver a computer system.

Part II

Development

Now that computer technology is widely used, the pioneering spirit of the first decades of computerization is giving way to a more professional organization of work. In addition to the task of developing and introducing new computer systems, computer professionals today have to spend much time maintaining and modifying old systems. But even if development is not what it used to be, computer professionals are still programmers who adapt computers to new and changing environments.

The task of developing computer systems is complex; there are no simple solutions. The number of theories, methods, and techniques dealing with how to develop computer systems is enormous. And we go on searching for better methods, experimenting with new approaches.

In this part we discuss three paradigms for development of computer systems, each of them providing a different perception of the task and the strategies to be used. These three paradigms are not to be understood as exclusive alternatives but rather as idealized viewpoints to be mixed in understanding and developing concrete approaches and projects.

The first two paradigms, construction and evolution, share the view that systems development is the solution to a given data processing problem. The construction approach concentrates on the complexity of the problem and suggests a rational and analytical strategy for the construction of a computer system. The evolution approach focuses on uncertainty and suggests an experimental strategy for problem solving.

In the third approach the problem is no longer given, and development is no longer seen as something isolated from the life of the organization. The relation between development and use of computer systems is stressed, and systems development is seen as an integral part of organizational change.

In part I we discussed three varieties of systems thinking. Here we discuss three approaches to development. How you think of systems will, of course, influence your thinking about development and vice versa. But there is no simple one-to-one mapping between the two.

4

Construction

Herbert Simon has drawn our attention to the fact that although the modern world is an artificial world, modern science is a science of nature. "The world we live in today is much more a man-made, or artificial, world than it is a natural world. Almost every element in our environment shows evidence of man's artifice. . . . Natural science is knowledge about natural objects and phenomena. We ask whether there cannot also be "artificial" science – knowledge about artificial objects and phenomena."

The natural sciences are of little help in making sense of our artificial world. They tempt us into taking technology and society for granted and prevent us from appreciating the role of technology in shaping society, its members, and their actions. Simon sets out to outline a program for a science of the artificial, and he immediately observes that such a science will be a science of design. If the world we live in is an artificial world, our understanding of the world must be an understanding of artifacts and how they are designed. Artifacts are put together by people; they are constructed. So, our first question will be, what does it mean to construct?

When an artifact has been constructed and put to use, people change it, adjusting it to their needs. Once an artifact leaves the hands of the professional constructor, it begins to evolve. But then, would it not be a good idea for the constructor to take part in this process, at least in the early stages of an artifact's evolution?

When we take the step from construction to evolution, we are ready to begin worrying about the actual use of our artifacts. It is not only the *artifact* that has to evolve – the *users* have to change in order to make the most of the artifact. Proceeding on this assumption, we involve ourselves in the introduction of our artifacts in the user organizations, and we intervene in order to change organizational structures and habits.

In this part we will discuss these different paradigms for computer artifact design – construction, evolution, and intervention – and in each case we will begin with a concrete example. In doing so, we are beginning to outline our response to Simon's request for a science of the artificial. Nevertheless, although we have used Simon as an inspiration, our discussion will go beyond his notion of design as decision making, and take us in a direction deviating from the quantitative approach he recommends.

Stepwise Refinement

Let's assume we have been given a normal 8×8 chessboard and eight queens that are hostile to each other. The problem to be solved is how to place all queens on the board so that no queen can be taken by any other queen. Our task as systems developers is to construct a program that will solve this problem, that will enable us to get from the initial state where the board is empty to the goal state where the eight queens are positioned in the manner specified.

To construct such a program, we must find a step-by-step procedure, an algorithm, for going from the initial state to the goal state. The program is to have access to a representation of the chessboard and operate by placing, moving, and removing queens on the board.

In the following, we shall draw on Niklaus Wirth's discussion of this so-called 8-Queens problem to illustrate how general problem solving techniques can be used, both in the approach to program development and in the program itself. In doing so, we shall describe stepwise refinement, one of the classical approaches to program construction.

Wirth first concentrates on defining a strategy for finding a solution, with the intention of implementing that strategy, and having the computer use it to find a way to the goal state. In considering how a person would go about placing the queens on the chessboard to achieve the goal state, he starts by proposing a brute force or exhaustive search approach: Go through all the possible configurations of queens until you either reach the goal state or have shown that it cannot be reached.

The problem with this strategy is that there are on the order of 2^{32} possible configurations. Exhaustive search is an example of pure bottom-up problem solving, in this case done mechanically by a computer. It is generally a good idea to begin in this way since it gives you a first feel for the problem. But with complex problems we do well to move on to top-down considerations trying to reduce the search space.

The 8-Queens problem *is* complex and Wirth therefore goes on to analyze the condition on the goal state into subconditions. He wants to find a way of using such subconditions to restrict the admissible operations of the program. The condition that no queen may be taken by any other queen can be analyzed into three subconditions: only one queen to each column, only one queen to each row, only one queen to each diagonal. The first of these subconditions will restrict the admissible operations to movements of queens within columns, one queen to each column. Then there will be only 2^{24} possible configurations, and we will need to test only rows and diagonals.

But we can do even better. There is not much more we can do by means of top-down analysis, so let us return to bottom-up considerations. Rather than trying out various configurations randomly, we can be systematic, ordering our trials so that the configurations are constructed in steps from other configurations, each step being much easier to construct than a whole configuration.

We then organize the search space of configurations to be tested into a tree in which we move from the root to the leaves, sometimes backtracking. Having placed a queen successfully in one column, we proceed to place one in the next column, and so on all the way to the eighth column, backing up to the previous column whenever we fail to place a queen in a column. Going through this procedure manually will, according to Wirth, take at the most 15 minutes. We have solved the problem, but as systems developers our work has just begun.

The data processing task of getting from the initial state to the goal state is now specified well enough for a human being to be able to perform it. But the specification is by no means detailed enough for a standard computer. We have found a strategy to solve the 8-Queens problem, but we are not finished with our programming task. We must proceed further to specify and write the code for a corresponding computer program.

Wanting to transform the solution of the 8-Queens problem into a computer program, Wirth now turns to the available programming facilities. Until we have done so, we are not really at the bottom of our programming task. As experienced programmers, we would certainly have been thinking all along about how to represent the board and the positions of the queens on the board and how to express our operations in our programming language. But we did well to postpone the introduction of programming terminology until we had moved from the original problem formulation to an informal specification of a solution procedure.

We begin the actual programming by analyzing the operations we want the computer to perform on the chessboard into suboperations until we reach operations that can be expressed in programming instructions. On the way we consider how best to represent the positions of the queens on the board, choosing a representation that optimally fits the instructions we are moving toward.

Since the success of the algorithm we have chosen depends on the choice of representation, we would do well to postpone this choice until we have a good idea of which instructions we will want to use. On the other hand, we cannot postpone the choice of representation too long, or we run the risk of not being able to find a suitable representation. In this way the design of computer systems involves skilled handling of complexity, of being able to keep track of interacting consequences between decisions.

Wirth uses the 8-Queens problem to illustrate and clarify the essential characteristics of stepwise refinement: A well-defined data processing problem is given, and the task of the systems developer is to construct a computer program for solving the problem. The development effort starts with an elaborate and systematic analysis of the given problem. Different problem-solving strategies are explored and a decision is made on which strategy to apply in the program.

The program is then constructed in a sequence of hierarchically ordered refinement steps. In each step a given operation is analyzed into a number of simpler operations. Each new operation defines a new refinement task for the systems developer. The modularity obtained in this way will determine the ease with which the program can be understood, modified, extended, or adapted to a changing environment.

The construction process is mainly top-down, even if implementation issues are considered in parallel. As systems developers, we rely on a notation natural to the problem at hand for as long as possible. At the same time we must keep an eye on the computer system and the programming language eventually to be used in order to determine the direction of stepwise refinement. Each refinement involves a number of decisions that must be made on the basis of design criteria such as efficiency, storage economy, and regularity of structure.

In summary, the method of stepwise refinement assumes that a well-defined problem is given. Following this approach, we start by analyzing the problem and deciding on a problem-solving strategy for the program. From there we continue in a stepwise fashion, using the evolving structure of the program to structure the process of constructing the program.

Think, Think, Think

As long as there have been people on earth, there has been fascination with systems construction. Think only of all the systems constructed by cosmologists in their attempts to make sense of the movements of the heavenly bodies. On the basis of incomplete, and often inadequate, data about the movements of the planets, together with presupposed principles of circular motion, people like Ptolemy (ca. 100–170) spent their lives constructing beautiful and complex cosmological systems.

Even if the accumulation of data over the years led to what seemed like a never-ending process of adding circles on circles to the Ptolemaic system, the cosmologists did not have to worry about changes in the reality they tried to represent. When Copernicus and later Kepler decided to change the fundamental principles of the Ptolemaic system, it was not because the planets had changed their ways and not mainly because they had better data. Rather it was because they had managed to break away from a conventional acceptance of the Ptolemaic principles.

To Ptolemy, it was evident that the heavenly bodies moved in circles, the circle being the perfect motion. It could not be otherwise. Kepler dared to abandon this principle only because he had become convinced that the general ellipse was even more perfect, the circle being a special case.

Ptolemy believed that there could be no change in the heavens. Likewise, eighteenth-century rationalists like Carl Linnaeus (1707–1778) did not believe in organic evolution. Individual organisms are born, live, and die, but the species remain unchanged, defined by their essential properties. Individuals of the same species differ only with respect to accidental properties. Linnaeus based his system for the classification of plants, his *Systema Naturae*, on the principle that reproduction is the essence of life and that therefore the properties of the sexual organs define a species.

These are examples of rationalistic systems constructions that lie rather far from the world of computer systems construction. But look at Wirth's example. The world of chess is an unchanging world. The principles determining the movements of the queens are simple and well-defined. We cannot argue with them and there is no room for alternative interpretations. Constructing our program, we are really closer to Ptolemy, Kepler, Linnaeus, and all the other great systems builders than we might think. And we have a lot to learn from them.

In this tradition, systems developers are rational thinkers solving abstract, complex problems. They are given information problems and they have computers that need to be programmed to solve the problems.

In performing their tasks, in bridging the conceptual gap between these two worlds, they concentrate on problems and possible solutions. They have to deal with large amounts of information, and to cope with this complexity they will use an analytical knife. Abstraction is used to focus attention and to select relevant information, and decomposition is used to structure understanding and create smaller and smaller subproblems. Each of the detailed subproblems are then solved and aggregated into the final solution.

The construction process is specification driven. The problem is described and analyzed in a requirements specification. Systems developers use specifications as means of remembering and communicating their thinking, and the task is seen as a transformation of specifications from a general problem-oriented level to a concrete machine-oriented level. In performing these transformations, systems developers use stepwise refinement in combination with structured techniques.

Structured analysis, structured design, and structured programming support systems developers in working in a disciplined way and in expressing their thoughts in a form that facilitates division of labor and leads to a robust and easy-to-maintain product.

Users play a somewhat passive role in such a construction process. They are at most objects in a process of requirements engineering, being interviewed in order to supply systems developers with a definition of the problem and the relevant information. All the creative thinking is done by the systems developers as they produce specifications and find solutions without actively involving the future users.

The role of the user is to provide information and approve decisions. The aim of the systems development effort is the production of a high quality system that meets the specified requirements. The result of the process is a computer system that is subsequently delivered to the users. The actual implementation of the system into an existing technological and organizational environment is not considered part of the development task.

By the way, who are the users in the 8-Queens problem? As you see, they are not mentioned at all. The problem is formulated in quite abstract terms, but then, of course, Wirth already knows about chess. The programmer and the user are one and the same person.

The idea that systems development is the construction of computer systems – using the method of stepwise refinement – has a dominating position in both theory and practice. It plays a major role in the traditional software engineering literature, and the project models of many software companies are based on it.

We have seen that systems construction is an idea deeply entrenched in our culture. The way this idea is expressed in software engineering is reminiscent of eighteenth-century rationalism, with its belief in rational thinking in a rational world and with mathematics as the exemplary science.

To say that the construction approach to systems development is rationalist is not to imply that other approaches are irrational. All the different approaches to systems development that we shall discuss are, of course, rational in the sense of giving reasons for their choice of approach. Each approach makes assumptions about the conditions and practices involved in systems development and is willing to defend these assumptions. But the fundamental assumption made in the construction approach is that systems development itself is a rational activity.

To be *rational* means, in short, to choose the optimal action given what we know and want. The rational mode of operation is to analyze and think, then to make a decision, and finally to act. The simplicity and power of this approach is obvious. By following this ideal pattern, we can economize with our resources and reach optimal solutions. Thinking beforehand of alternatives is quicker and more prudent than trying them out in practice; and imagining a catastrophe is better than experiencing one. The true rational practitioner is an economic man.

The other side of the coin is that successful rational action requires that a number of conditions are fulfilled. The goal or problem has to be stable and explicitly stated. We must have information about alternative ways to reach the goal or solve the problem. We must be able to measure and compare the consequences of each option. And, finally, we must be committed to solve the problem and have the resources necessary to do so.

If one or more of these conditions fail, we shall have to supplement our rational approach with other compensating activities. If the problem is changed when a solution is already found, then we must accept working with a solution that no longer fits the problem – or we shall have to begin all over again. If we don't have enough resources and enough time to analyze all relevant options, we must accept satisfactory rather than optimal solutions – or we must engage in administrative and political activities to get more resources.

Abstract Problems and General Methods

In stepwise refinement the construction of a computer system is analogous to the construction of a proof in mathematics. And the construction of a proof is understood as the solving of a problem.

Problem solving has become one of the major research areas within contemporary cognitive science. A theory of problem solving is almost a general theory of thinking. Psychological research, in combination with the reports of experienced problem solvers, has given us a standard view of what a problem is, as well as some general truths about how human beings solve problems.

A problem is defined by two states: the initial state (where you are) and the goal state (where you want to be). That is the abstract formulation of a problem. To solve it, you will have to find a route (Greek: *methodos*) leading from the initial state to the goal state. You search through a space of possible solutions looking for a route from the initial state to the goal state. Finding the goal state by chance, without knowing how you got there, does *not* mean you have solved the problem. The solution is not the goal state but the method, the way to get there.

There are two types of general problem-solving methods, two different ways you can go about solving a problem. Unless the problem space is too large, you can always perform a complete search, using more or less elegant strategies. If for some reason you cannot or do not want to do a complete search, you must reduce the problem space, using heuristics to guess your way through. General problem-solving methods help you find solutions to problems – that is, they help you find your way from problem to solution. The solution is itself a method, a route that leads from the initial state of the problem to its goal state.

Human beings solve problems using means-ends analysis, ultimately showing how the means available can be put together to attain the end. They work by combining a top-down approach, breaking down the problem into subproblems, with a bottom-up approach, playing around with the available means.

These general ideas about human problem solving appear in many different concrete forms within different forms of human activity. Mathematics and sociology each have their own traditions and debates about problem solving. The same holds for medicine and law.

Systems development is, in this context, a young discipline that has been influenced by general problem-solving ideas drawn from different sciences and disciplines. One way to understand better the different approaches to programming and systems development is to focus on some of the basic questions: How are problems formulated and interpreted? How are methods related to problems? How are complete search and heuristics, top-down and bottom-up approaches, used and combined?

Specific ideas and methods for solving mathematical problems have played an important role in shaping our ideas about programming and systems development. On one level, such problem-solving methods are used in the design of programs, when solving data processing problems. They are used as methods for identifying algorithms and organizing data. We recognize the strong influence of such mathematical approaches within various areas of computing – for example, sorting and searching – and in relation to computer applications such as compilers and interpreters.

On a more general level, such methods are used to organize the processes of programming and systems development. They become methods of systems development. This kind of influence is apparent in techniques for formal specification of programs and in methods for program verification.

The system construction approach is in many respects an expression of mathematical problem-solving ideals. The starting point is a precise formulation of an abstract problem. The problem is known and it is not open to interpretation. Also, a number of general methods are known, and we apply these to solve the problem. The problem and the methods are known. The challenge is to intelligently apply the methods to produce a solution. It is no coincidence that Wirth chooses a chess problem, in effect a mathematical problem, to illustrate his views on software engineering. The world of chess problems has all the characteristics of a mathematical system: it is stable, closed, well-ordered, and knowable.

The construction approach relies on hard systems thinking. A computer application is conceived of as a system. The systems developer operates on the representation of a future computer application on the assumption that eventually the representation will be used as a blueprint for a real system.

The computer application is thought of as a system of hierarchically ordered subsystems. The structure of the system is used as a framework for organizing and managing the construction process. From one point of view, modules can be understood as parts of the system. From another point of view, they are simply specifications of work tasks to be performed. The modules of a structure chart are specifications of well-defined programming assignments that can be performed by different programmers. Narrow and well-defined interfaces between modules make it possible to implement each module with a minimal amount of coordination between different programmers.

In this way the hard systems approach is used both for dealing with problems and solutions *and* for controlling the process of constructing the system. Methods for systems construction are based on hard systems thinking and they provide us with concepts and metaphors for understanding both products and processes. The construction process is seen as a machine, with its different functions reflecting the functional parts of another machine, which is the computer system to be.

The idea of software development as construction, as mathematical problem solving by stepwise refinement, defines the discipline of systems development for the majority of practitioners and researchers. It has become a *paradigm* in Thomas Kuhn's sense, unifying the community, giving it a common identity. Within the general confines of this common understanding, there is an abundance of rivalling methods, but the discussion of the pros and cons of these methods can presuppose a common background of fundamental beliefs and values.

In such a paradigm there are beliefs about the subject matter of software development, what it is and some general truths about it, as well as beliefs about knowledge and the methods by which it is obtained. But most important are the *exemplars*, model examples of how software development is done when it is done well. These exemplars can be real or invented. They are used in teaching, and they can be used when we cannot spell out in explicit detail what we do as systems developers. We learn the methods by studying the examples, and we remember the methods by remembering the examples.

The details of the exemplars can differ within the community, but their typical properties must be the same. The 8-Queens problem is a good candidate for such an exemplar when we are talking about systems development as system construction. Its subject matter is well-defined data processing problems, the solutions of which are to be turned into computer systems. The problems are abstractly defined, and there are no users involved. Both the problems and their solutions have a hierarchical organization and admit of modularization. We move around in what is basically a mathematical world, using hard systems thinking.

Some would argue, that the 8-queens problem has little or nothing to do with real-world problems, just as the development of computer systems has nothing to do with solving chess puzzles. It is a completely different world. Literally speaking that might well be so. But how do we approach the so-called real problems? How do we think about them? What kinds of methods do we apply? Accounting, production management, and personnel administration are definitely not chess. Still, our discipline is deeply influenced by concepts and methods that are as far from accounting, production management, and personnel administration as chess is.

The 8-queens problem is a good candidate for a paradigmatic exemplar because it so clearly shows us the strengths, but also the limitations, of one of the most powerful paradigms within our discipline.

It's All Routine

Within the systems construction paradigm, practitioners are more like heart surgeons than general practitioners. Just as the heart surgeon has to realize that a pacemaker or a bypass operation won't solve all health problems, systems developers know that computers aren't the solution to every organizational problem. Therefore, they deal only with explicitly stated data processing problems to which they apply a set of standard techniques to find an optimal computer solution.

Users and clients are themselves responsible for issues related to their organizations, and they identify and formulate data processing tasks. They know their business and their professions. They have the competence needed to find and make explicit relevant areas in which to use computers. Systems developers are experts in applying computers to meet data processing needs. They concentrate on constructing computer systems in response to the problems given them by their clients. Like the heart surgeons, their work begins when the diagnosis is made: A data processing task has been formulated, the computer has been identified as a feasible technology, and the challenge is to construct a computer system as a solution to the problem at hand.

Heart surgery takes a lot of competence, and the required skills change as new techniques are developed. But each individual heart operation is pretty much routine work. The cases are similar and the surgeon applies state-of-the-art techniques provided by research and development efforts. The excitement comes from the practical difficulties, the concentration needed and the importance of the task. As a surgeon, you are literally holding a person's life in your hands.

The construction of computer systems is largely a matter of routine as well. Here the excitement comes from the complexity of the task, the prestige of working with high technology, and the impression of being at the heart of progress. Like the surgeon, systems developers apply state-of-the-art techniques to cope with the complexity involved in constructing computer systems, and they gradually expand their repertoire of methods and techniques as new and more effective technologies are developed.

Construction is a bureaucratic approach to systems development. Chief programmer teams, phase models, documentation standards, and structured techniques are all examples of bureaucratic techniques. The traditional sequential life cycle model divides the construction process into distinct functions to be performed in separate time intervals or by different persons. First survey, then analysis, then overall design, and so on. Methods are applied in performing each of these tasks. These methods provide techniques to guide the actions of the systems developer, they provide a set of tools to formulate and manipulate specifications, and they suggest how to organize cooperation between the involved actors.

The problem is formulated together with an overall plan as a contractual arrangement with the client organization in the early stages of the process. Such a project charter is a one-shot contract fixing the total amount of resources and time to be spent on the project. The plan is a result of negotiations between the involved organizations, and it is based on expert estimates of the effort required to solve the given problem.

Life cycle models and methods are general prescriptions for the construction process. On a specific level, they are supplemented by the project plans. Project plans are concrete implementations of the general guidelines, expressing how the specific system is going to be constructed. The overall plan is detailed and it is elaborated on as the process develops and a solution is designed. Project plans and specifications of the system are opposite sides of the same coin. The specifications describe the tasks to be performed, and the project plans describe how these tasks are distributed as work assignments over time.

The idea of a bureaucratic approach is, as we know, to prevent direct interaction between the divided tasks. Coordination is achieved by having each group or individual follow general company standards, rules, and specific instructions expressed in project plans and design documents. If these prescriptions do not apply, explicit coordination is needed between the involved actors, which may result, for example, in the redesign of the interface between two modules. We also know that the weakness of the bureaucratic approach is its inability to respond effectively to a changing environment.

Systems construction is mathematical problem solving. But the problems we try to solve when constructing computer systems are often mathematically trivial, and our solutions are far from elegant. Any task can be viewed as a problem, and computer technology turns trivial tasks into new problems. Improvement is often made possible by the use of new technology, and thus computer technology turns simple information tasks into data processing problems to be solved by computer systems.

But when simple routine tasks are turned into problems to be solved by a computer program, indicating a preference for brute force over elegance, the mathematician will begin to complain: "Those are not problems at all, and those brute force searches are not solutions either." What Wirth is doing is too mechanical, almost bureaucratic, and mathematicians are no bureaucrats. They count only those solutions that have at least a minimum of elegance. They solve problems in order to learn about the problem area. They are not just interested in reaching the goal state. They want the solution to tell us something, to point to some important property of the problem space. If there is no such solution, the problem is not really worth spending time on.

Wirth never really tries to solve the 8-Queens problem, the mathematician would say. Wirth wants to find a general method, suitable for developing computer systems, for solving these kinds of problems. That is his problem; that is not the 8-Queens problem.

To solve the 8-Queens problem, we would, of course, try to use the symmetry of the chessboard. Maybe the problem could be solved in steps by somehow reducing the board? Along the

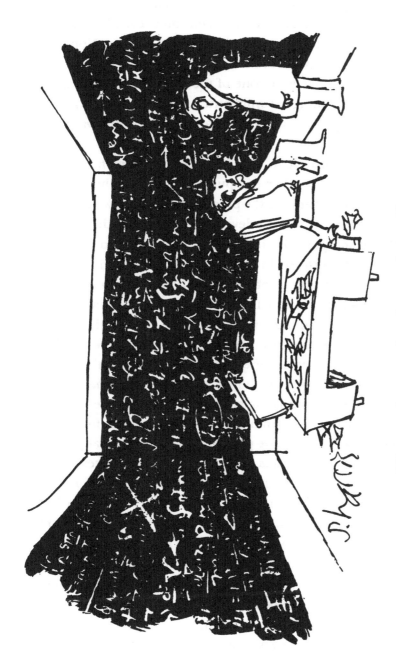

"Whatever happened to *elegant* solutions?"

diagonals? What if we solve the problem for a 4 × 4 board? Could we then find an easy way to expand the solution into one for an 8 × 8 board? Can we place four queens on a 4 × 4 board so that they don't threaten one another? That is easy. Notice the symmetry. Now, expand the board by dividing each square into four, placing the queen on one of the four squares, adding four queens, keeping things as symmetric as possible. Well, this is not really an elegant solution, but perhaps we are on our way toward one. What do you say?

Wirth's approach to the 8-Queens problem has, in fact, been challenged by other computer scientists. In his introduction, before embarking on the presentation of his own approach, Wirth invites the reader to develop a solution to the problem. Peter Naur accepted the invitation and decided to record his problem-solving process in a detailed diary. Naur documented his thinking about the problem in 36 notes, and he solved the problem in approximately 15 hours spread out over a period of one month, while he was engaged in many other activities.

In comparing his experience with Wirth's approach, Naur documents a number of differences. Whereas Wirth's program produced 92 solutions, including all symmetric variants, Naur spent some effort eliminating symmetric solutions. He identified the symmetry problem from the outset, without at first having any idea how to solve it. The problem continued to reappear and was not settled until he was halfway through the process.

Several times Naur pursued alternatives in parallel, many of which were later rejected. Some decisions were made in large steps, each including many issues and alternatives, in contrast with the piecemeal approach of stepwise refinement.

Timing and efficiency considerations were explicit and quantitative, and they played a role throughout the entire process. And finally, the possibility of logical errors made by the programmer was taken into account.

Drawing from his own experience of solving the 8-Queens problem, Naur concludes: "At least some program development does not, and can hardly be made to proceed as the top-down process advocated by Wirth. Rather, a problem solving type of process is taking place. In this process high-level program descriptions certainly appear, but only after important details have been explored at a lower level."

The systems construction paradigm is a bureaucratized version of mathematical problem solving, applied to the commercial business of installing new technical equipment in workplaces. As a methodology for programming under mathematical conditions, this idea is fine except for its underestimation of the value of making errors. It is a truism that people learn from their mistakes; but this truism rarely enters the world of scientific and technological methodology.

The scientific method, for example, is generally taught as if hypothesis testing is a matter of finding out if a hypothesis is true or false. More realistically, what we do is to try to learn more about a hypothesis by finding out *how* it is false. Similarly, good mathematicians, problem solvers, or programmers savor their errors as outstanding opportunities for learning just when and how their ideas don't work.

As a paradigm for systems development, the idea of construction is, of course, incomplete. In real systems development projects, the conditions (criteria of success) of problems are undecided or vague, the admissible operations are not specified, and there are several viewpoints resulting in different sets of admissible operations and different criteria of success.

Problem solving is often a matter of creative discoveries of possible operations, perhaps by means of seeing the problem differently, from a different viewpoint, by making a creative recategorization of the problem. When a user says in the middle of a development effort "I would like such a system instead," we are confronted with a problem that cannot be solved by stepwise refinement.

5

Evolution

Specifications For Prototypes
The Problem Is Alive
Complexity and Uncertainty
Little by Little

If you happen to be on a certain South Pacific island on the right early spring day, you may come across a disturbing spectacle: From millions of turtle eggs laid in the grass some 40 yards from the edge of the sea, millions of tiny black turtles are hatched within a few hours. As soon as the turtles are hatched, they begin a perilous race across the white beach to reach the relatively safe ocean. Thousands of sea gulls descend on the white beach, devouring the newborn black turtles by the millions. Only a fraction of a percent of the turtles make it to the ocean. What a cruel waste!

The unfortunate turtles illustrate Charles Darwin's theory of evolution by natural selection. Darwin's ingenious explanation of how evolution proceeds by blindly trying out new life forms, only to see most of them extinguished, is difficult to fully accept. Even if errors are great sources for learning, the process of trial and error seems brutally stupid. But that is nature's way.

Foresight and planning resulting in the single, correct, final solution seems so much more sensible. But planning is a risky business. What works wonderfully on paper often fails miserably in reality. When programs gain complexity, we need to run them to make sure that they really do what they are supposed to do. This is so even if we assume the technical and organizational environment to be stable and reliable. And we know that this assumption is unwarranted. Computer technology is developing, and we are often working with untested material. Organizations compete with one another and embody wide spectra of conflicting interests resulting in changing plans and strategies.

The 8-Queens problem was solved by using well-known techniques. The problem is complex, but its specification is clear and exact. Real data processing problems in organizations are seldom like that. Real problems are difficult to solve because they are unclear and elusive, and because they change.

In real projects the challenge is not only complexity but the equally important uncertainty related to the data processing problem. This suggests an experimental rather than an analytical approach. Errors are not to be viewed as mistakes. Instead, trial and error is a constructive approach to reduce the uncertainty. Errors are sources for learning.

This fundamental rationale of an evolution approach is expressed well by Piet Hein:

> The road to wisdom?
> Well, it's plain and simple to express:
> Err and err and err again
> but less and less and less.

In the following section we shall report on a programming experiment conducted by Barry Boehm. The experiment which compares the evolution approach with the construction approach illustrates the strengths and weaknesses of each.

Specifications or Prototypes

The task is to develop a single-user, interactive system to be used by project managers for estimating the resources needed in a systems development project. The estimation procedure should be based on Boehm's specific technique, which he calls the Constructive Cost Model, or Cocomo. The system, hereafter called a Cocomo system, should be easy to use for an experienced but impatient project manager who is acquainted with this estimation technique. More specifically, the system is required to be robust, easy to use, easy to learn, and functionally equivalent to the Cocomo technique.

Cocomo is an empirical cost estimation model that uses data from previous projects to derive an estimate on the current project. A simple background equation is used to derive a rough estimate of the total effort in man-months as a function of the size in thousands of lines of delivered source code. Sixteen different factors – such as product complexity, main storage constraints, analyst capability, programming language experience, and applications experience – are then analyzed and evaluated based on previous experiences. The final estimate is derived by adjusting the first rough estimate accordingly.

In the programming experiment, several groups of graduate students in computer science were engaged to test different approaches for developing a Cocomo system. Four groups were asked to develop systems using an analytical approach based on specifications, what we have called a construction approach. Three other groups were asked to develop their Cocomo systems by using an experimental approach based on prototyping.

The specification groups began by developing a requirements specification and later an overall design specification. Both specifications were subjected to thorough reviews before the technical design, programming, and implementation of the system were begun.

The prototyping groups began by developing a prototype. The prototype was evaluated by reviewers, who provided feedback on errors, pointed out shortcomings, and suggested modifications. The evaluated prototypes were then used as a basis for implementation.

All the groups were given similar conditions for developing the system. Each group was expected to create a comprehensive documentation of the development process, and afterward each of the resulting systems were evaluated by external reviewers.

Now, what can we learn from this experiment? The idea is, of course, to learn something about the relative advantages and disadvantages of different approaches to software development. We would like to know how the chosen approach affects the development process – for example, the cooperation between the involved actors or the amount of resources needed to develop the system. We would also like to know about differences, resulting from the choice of approach, on the quality of the system, in this case with respect to functionality, robustness, ease of use, and ease of learning.

In practical systems development, every project is different. We never go through the same process twice, and we rarely switch back and forth between general approaches. When we change our ways of working, it is for reasons of fashion or because we have to adjust to a new company style. So we never really test our most basic assumptions.

An experiment like the one described above provides us with such an opportunity. The problem with the experiment is the artificial nature of the involved development projects, and the uncertainties related to the comparisons between the individual projects. But this seems to be a fundamental condition for research and development in our field.

A major result in Boehm's experiment was that prototyping really did seem to have a number of advantages as compared to a specification approach. Prototyping resulted in systems with roughly equal performance, but requiring less than half the code. The prototyped systems were rated higher in their ease of use and ease of learning. They had better human-machine interfaces. The prototyping groups had the advantage of having a functioning system very early in the development process. There was a reduced deadline effect toward the end of the prototyping projects.

The other major result was that these advantages were not for free. Proportionally less effort was spent on planning and designing, and proportionally more effort on testing and fixing. The integration of subsystems was more difficult because there was lack of interface specifications. The resulting design was less coherent. In fact, the prototyped systems were rated lower in overall functionality and in their tolerance of erroneous input – that is, in robustness.

With these differences in mind, the overall performance of the systems was the same for the two approaches tested. In addition, the experiment did not reveal any significant differences, in productivity in terms of delivered source instructions per manhour, between the two approaches.

The results of the experiment raise several questions. How can we account for the differences between the two approaches? And what practical consequences should these findings have for systems development work? Let us begin by discussing the results in relation to a more general understanding of an evolution approach to systems development.

Why did the two approaches lead to systems with different qualities? Intuitively, we should expect that prototypes help us design the human-computer interface, emphasizing properties like ease of use and ease of learning. A prototype illustrates the outer form of the system, the user interface is made explicit, and we can evaluate it by trying it out. Similarly, our intuition tells us that an analytical, systems construction approach helps us understand the given problem and design a solution in accordance with functional requirements. The differences between the qualities of the resulting systems are not surprising.

The Problem Is Alive

We can give a less superficial interpretation of Boehm's results, if we take into account the nature of the given problem. Like the 8-Queens problem, the Cocomo problem is in many respects well defined. A key feature of the problem formulation is that the system should be based on an established estimation procedure. Since we have agreed to use a definite procedure for calculating the estimate, we have, in effect, decided what kind of data the system will use.

But in contrast to the 8-Queens problem, the formulation of

the Cocomo problem includes project managers as future users of the system. The problem is no longer an abstract data processing task. Instead, the future context for using the system is included in the problem formulation. The problem has a human dimension.

The Cocomo problem includes requirements of a different nature than the ones found in the 8-Queens problem. How do we decide whether a system is easy to use or easy to learn? To what extent are we able to design such qualities as a result of a thorough analysis? How dependent are we on the idiosyncratic preferences and customs of the users?

In a construction approach, the challenge is to develop a system in accordance with the specified requirements. The success or failure of the system is determined by comparing the problem formulation with the resulting system. This comparison is made intellectually through formal argumentation and practically through various kinds of tests.

In an evolution approach, the problem is also given and the criterion of success is basically the same. But other kinds of qualities are included. The context of the system is included together with the individual interpretations of different users. As a consequence, the issue of measurement and evaluation has become problematic.

Evolution approaches play a dominant role in the literature, and they are widely used in practice. These approaches are, like the systems construction approaches, based on the assumption that a problem is given and that the task is to develop a computer system in response to this problem. But in contrast to the construction approaches they recognize and emphasize the uncertainties related to the specific problem in question and to systems development in general.

The process is bounded by limited knowledge and limited resources, and the systems developers have limited capacity for developing new and useful insights. Some alternatives are not or cannot be recognized, and the consequences of choosing an alternative at a specific point of development cannot be fully understood. What will the response time be if we choose this design as opposed to that one? Will the system be easier to use if we design the interface in this way? These questions are difficult to answer on a rational basis before we have the opportunity to test and use the system. And even then we are facing nontrivial problems of evaluation and measurement.

With an evolution approach one will argue, with Herbert Simon, that the best we can hope for is bounded rationality – that is, systems developers who exercise "satisficing" rather than optimizing behavior. The result of a development effort is not the final solution to a given problem. Instead, the problem has to be interpreted and restated, and the result is a satisfactory version of a system that might be developed further, based on the practical experience of using it.

In the evolution approach, systems developers are scientific investigators rather than economic men. Basically, they process information as they move along. Interpretations and decisions are made without knowing for sure whether they will prove useful or not. Problem formulations, possible solutions, and decisions are of a hypothetical nature. Some of them are rejected later whereas others are pursued further. Prototypes and other kinds of experimental artifacts are used intensively. Systems developers spend more time identifying and experimenting with possible solutions than they do analyzing problems.

Systems constructors favor a top-down approach. They develop systems in a stepwise fashion, decomposing and refining an abstract conception of the system as a whole into its concrete components. In contrast, evolutionists use a bottom-up approach. They identify and evaluate concrete solutions to partial problems, gradually approaching the system as a whole.

Systems developers that use an evolution approach are strongly dependent on problem owners and future users. Users play an active role in evaluating design proposals and prototypes, and problem owners are needed to negotiate and make decisions regarding the problem formulation and the quality of the produced system.

Early experiments in transportation

Quality control is not simply a matter of evaluating systems and specifications in relation to given problems and requirements. Somehow, we have to decide which alternatives to reject and which to pursue, and in the end we have to decide whether a proposed solution is satisfactory.

In the construction approach, systems developers are technical experts who on their own are able to find the best solution to a given data processing problem. In the evolution approach, systems developers are still technical experts but rather than being able to construct the best solution by themselves, they have to become teachers and facilitators as well. They propose and develop technical solutions; but throughout the development process, they communicate and interact with problem owners and users. They engage them in evaluating alternatives and making decisions.

In addition to being technically competent, the systems developer must be able to explain and discuss problem formulations and technical solutions with users and problem owners. The systems developer is both a technical and a communicative expert.

Complexity and Uncertainty

The Cocomo experiment illustrates the major differences between a construction approach and an evolution approach. Even if we have discussed these differences, we have not answered the most important question: What kind of practical consequences should the results of Boehm's experiment have for systems development?

One way to approach this question is to distinguish between types of situations. We can look at construction and evolution as two alternative approaches, and then reformulate the question as follows: In which situations should we choose which approach?

The main concern in the construction approach is complexity, whereas the main concern in the evolution approach is uncertainty. This suggests the following principle: In situations where the complexity of the problem is high and the uncertainty is low, choose a construction approach; in situations with high uncertainty and low complexity, choose an evolution approach.

This sounds reasonable. But what if both the complexity and the uncertainty are high? Which approach should we then choose? Maybe we are wrong in looking at the two approaches as alternatives? Maybe we shouldn't ask which one to choose, but rather how to combine them.

Wirth is arguing in favor of a construction approach, and Naur, in his criticism of Wirth, is arguing for an evolution approach. Wirth is in favor of a rational top-down approach based on deductive, stepwise refinements of specifications. In contrast, Naur is in favor of an experimental bottom-up approach based on trial and error and the idea of developing a solution out of concrete solutions to partial problems.

Naur refers to two general problem-solving principles. The first principle – "Run over the elements of the problem in rapid succession several times, until a pattern emerges which encompasses all these elements simultaneously" – emphasizes the bottom-up nature of the evolution approach against the top-down strategy of the construction approach. The second principle – "Suspend judgment. Don't jump to conclusions" – points to the openness and uncertainty of the problem.

It is worth noting that Wirth and Naur argue from the very same example. Both of them have developed a computer program to solve the 8-Queens problem. The only difference of importance between the two situations is the programmer. Now, should we argue that one of them is wrong, or should we argue that the choice between a construction approach and an evolution approach is a matter of personal style and preference? Neither of these options seems very attractive. Both approaches and the arguments for them are appealing. And even if individual differences are important, our profession needs to offer more constructive advice on how to improve practice. Analyzing the argument between Wirth and Naur, it seems fair to assume that the two approaches should be combined.

We should also keep in mind the results of the Cocomo experiment. The prototyped systems had better human-machine interfaces. They were rated higher in their ease of use and ease of learning. But the resulting design was less coherent. The prototyped systems were rated lower in overall functionality and robustness when compared to the systems that were developed using a rational approach based on specifications.

Functionality, robustness, ease of use, and ease of learning are hardly alternative quality characteristics. In most cases we would like computer systems to possess most, if not all, of these properties. Again, this is an indication that we should understand construction and evolution, not as alternatives, but as complementary approaches to systems development.

The question of complexity versus uncertainty is one important key to understanding the relation between the two approaches. The degree of complexity in a given situation is a measure of the amount and diversity of relevant information needed to solve the problem. The more diverse the information is, the higher the complexity. In contrast, the degree of uncertainty represents the accessibility and reliability of information that is relevant in a given situation. The more accessible and the more reliable the information is, the lower the uncertainty.

Both in construction and evolution we attempt to reduce the problem or task by applying a specific information-processing strategy. In the systems construction approach we want to reduce the complexity by using abstraction and decomposition. When we face too much information, abstraction helps us focus our attention by neglecting certain aspects and concentrating on others. Decomposition helps us move from an abstract overall understanding toward a more concrete one, still using our abstract understanding as a frame of reference and thereby maintaining a manageable level of complexity.

In the evolution approach we attempt to reduce the uncertainty by using trial and error, by interacting with the environment. When we face uncertainty, we need more information to learn about the unknown and to evaluate what we assume or believe.

From this information-processing point of view, both approaches seem far too optimistic in what they promise. If we behave in a rational way based on abstraction and decomposition, we rely on a simplified world, and in doing so we introduce new uncertainties as to what extent this view is in accordance with the complex real world. Correspondingly, when we behave in an experimental way we produce new information as we go along, thereby introducing more complexity. Complexity and uncertainty are intrinsically related aspects of problem-solving situations.

If we rely on a rational approach, we introduce uncertainties requiring experimental countermeasures. And if we rely on an experimental approach, we introduce complexities requiring rational countermeasures. This *principle of limited reduction* in designing solutions to given data processing problems suggests that effective systems development requires a combination of rational and experimental approaches.

Practical models on how to combine construction and evolution approaches have been suggested as alternatives to the classical waterfall model. One of the early models is the so-called spiral model. The basic idea is to analyze and evaluate the risks related to a project, to resolve these risks by using a wide range of rational, analytical, and experimental techniques, and then to use these insights as a basis for specifying and implementing the computer system.

The process evolves in learning cycles, each cycle encompassing risk analysis, risk resolution, and traditional specification activities. As the risks are reduced to an acceptable level, the project continues in a more traditional fashion following the waterfall model.

The spiral model is a practical way to implement the principle of limited reduction. The model effectively combines the strength of an organic and open mode of operation in the early stages of a project with the strength of a bureaucratic and production-oriented mode of operation in the later stages.

Due to the very nature of the evolution approach, it is difficult to plan and design the development process in advance. The project structure reflects the learning process rather than the structure of the system, and it is designed and developed in parallel with the system. Proportionally less effort is spent on planning the process and designing the system, and proportionally more effort is spent on fixing the process and testing and modifying the system. This was obvious in the Cocomo experiment.

The evolution approach relies on an organic approach in managing the development project. There are few permanent rules and structures prescribing or determining the behavior of the actors. To achieve coordination, each project has to establish and dynamically maintain explicit interaction between all parties. Important information related to organizing the development effort will be revealed only as the activity is performed. As a consequence, it is more difficult to give a reliable estimate for a project based on an evolution approach than for one based on a construction approach. For instance, using Cocomo guidelines, we have to predict the size of the program, but the very idea of experiments is to develop new and different conceptions of the program. How can we then, before the experiments, provide reliable estimates of the size of the program?

We should expect that the evolution approach requires that systems developers spend more energy on project management to effectively cope with the high task uncertainty. In the construction approach, there is much emphasis on planning the process and designing the system before action is taken. But relatively little effort is then spent on adjusting and modifying plans as the process unfolds. In the evolution approach, there is less emphasis on planning and design. But during the process, considerable time must be spent on establishing cooperation, on explicit coordination, on evaluating the status of the project, and on adjusting and further developing plans and patterns of cooperation.

The evolution approach assumes the environment to be dynamic in the sense of being vaguely understood or even unpredictable. The problem is, so to speak, alive. Under these conditions, an organic project organization is more effective than a bureaucratic one.

The very point of the evolution approach is that systems developers have limited knowledge about the environment of the process and that the environment contains information that is highly relevant to solve the problem. To behave rationally under these conditions, systems developers are forced to go beyond a simple rational approach. They have to interact with the environment, accept the openness of the problem and the system to be developed, take into account the preferences and beliefs of problem owners and users, deal with the economical and political climate of the project, and keep in step with the changes in the kind of technologies on which the project is dependent.

If the way Wirth handles the 8-Queens problem will serve well as a paradigmatic example of systems construction, then the Cocomo problem is a good candidate to serve as a paradigm for the evolution approach. The complexity of the problem is quite low because the system is to be based on the Cocomo model for estimating software projects. But the uncertainty related to the design of the interface and the reaction of the users is quite high. The problem invites an evolution approach.

In Scandinavia an important real-world project of the early 1980s has served as a paradigmatic example of the evolution approach, emphasizing the interaction between systems developers, problem owners, and users. In this project, called Utopia, systems developers worked with graphic workers to develop a workstation based on their professional expertise. The actual construction of the computer system in Utopia involved a lot of technical maneuvering, but the solution to the problem was found by stepping into the world of graphical work and newspaper production, and by emphasizing a strong interdependence and interaction between systems developers and graphic workers.

Little by Little

The dispute between top-down systems constructors like Niklaus Wirth and more bottom-up oriented problem solvers like Peter Naur is not specifically related to the development of computer systems. Most every science or discipline has its Wirths and its Naurs.

Carl Linnaeus is, as we have seen, an outstanding example of a systems constructionist in biology, deducing his taxonomic system of plant identification from general principles, presupposing a stable, unchanging system of nature. But in biology we also have someone like Charles Darwin with his ideas about organic evolution, of a nature constantly changing, little by little, by "imperceptibly small steps." While Linnaeus believed in eternally unchanging species defined by their essences, Darwin argued that individual organisms are importantly different even within what we normally think of as one species. While Linnaeus believed that biology was very much a question of logical reasoning, Darwin was a firm believer in observation, in the painstaking collection of data from nature.

These fundamental differences between Linnaeus and Darwin are often summarized by saying that Linnaeus was a rationalist while Darwin was an empiricist. Let us try to deepen our understanding of the construction and evolution approaches by looking closer at rationalism and empiricism.

The mechanistic world view was given its definite shape by scientists like Descartes and Newton in the seventeenth century. Descartes laid the groundwork for what was to be called rationalism, a belief in the rational order of the world with mathematics as the supreme example of human knowledge. Newton inspired his fellow British philosophers to formulate empiricism, a theory of knowledge that took physics as its model and believed in sensory experience as our main source of knowledge.

The rationalists became system builders, utopian society designers with a belief in final solutions and bureaucratic organizations. Social problems were like mathematical problems: They could be solved once and for all if only the adequate principles for the social order were found.

The empiricists became social engineers, relying on empirical regularities to explain and control society. They often cautioned against utopian design, advocating piecemeal change and small scale experimentation to produce major social reforms. But at the same time they were inspired by Newton's system of mechanics, which they believed to be the definitive and complete physical theory. In the end they thus often tended to be utopian believers in final solutions, as were their rationalist adversaries. Like the rationalists, the empiricists believed in a well-ordered world, governed by a few fundamental principles.

The construction approach belongs to the rationalist tradition while the evolution approach is an expression of more empiricist ideals. Constructionists and evolutionists both share a mechanistic world view. The difference between them lies more in their attitude toward knowledge, what they think they need to know in order to solve their problems, than in their basic assumptions about the world.

But is that right? Doesn't the evolution paradigm practice soft systems thinking and thus belong within a romantic world view? There is no simple answer to these questions. As we have described the evolution paradigm, it counts among its members both Herbert Simon and Peter Checkland. Within the evolution paradigm, there is room for mechanistic, hard systems thinking as well as for romantic, soft systems thinking. The decisive move from construction to evolution is when you give up deductive rationalism for an empiricist, inductive, experimental approach.

Deduction is the method of mathematics. Theorems are proved by deducing them from already established truths. The deductive method was immensely popular among the rationalists in seventeenth- and eighteenth-century Europe. Inspired by *The Elements,* Euclid's (ca. 300 B.C.) impressive treatise on geometry, the rationalists dreamed of arranging all knowledge in an axiomatic system, deducing every truth from a few fundamental axioms. When we want to use a requirements specification as the sole basis for the construction of a computer system, we dream this rationalist dream of deducing the program from the specification.

While Descartes was trying to deduce all truth from a single indubitable axiom, "Cogito" ("I think"), and Leibniz was putting together the key principles for his system of the world, philosophers and scientists in Great Britain were developing a very different view of knowledge. Rather than deducing the truths about the world from unarguable axioms, these early empiricists claimed such truths should be collected by unbiased observation. From large collections of individual data a general principle could be inductively inferred, they argued. They were ardent proponents of this method of induction, but the method itself was seen to be problematic almost from the very beginning. How can we infer a general principle from a finite number of observations?

Notice the parallel between these theories of knowledge and the theories of politics so fiercely discussed in those days. Rationalism in politics dreams of turning society into a hierarchically organized axiomatic system based on fundamental principles. It is the idea of a well-organized, well-planned, bureaucratic system based on fixed, legal rules. This was the kind of society designed by the French revolutionaries. Empiricism, on the contrary, leads to a liberal or anarchistic view of society, a democracy in which individuals contribute "data" to the government of a loosely organized, changing network of people.

The fierce antagonism between rationalists and empiricists has mellowed over the years. In this century, science (and politics) has arrived at a somewhat loose but healthy combination of the two methods. The kind of inductivism stressing the need for unbiased observation has been abandoned as impossible in favor of a method of theoretically biased hypothesis testing. We use deduction to infer the empirical implications of our theories and then we test these implications by collecting data, using induction to order the data.

It is this healthy mixture of deduction and induction that we would like to see in systems development, combining systems construction and evolution ideas. But then Wirth's account of how to solve the 8-Queens problem cannot be used in isolation as a model for how to do systems development. There are elements of bottom-up thinking in Wirth, to be sure, but there is no acknowledgment of a world outside the problem that might provide information relevant to its solution. Corresponding objections can be brought against Peter Naur and the Utopia project. Both Naur's two principles quoted above about suspending judgment, and the extreme reliance in Utopia on the users really seem to be expressions of extreme inductivism. But perhaps we can read Naur as giving us only the very sensible advice to stay cool and look before we leap. And in the Utopia case the inductivism is perhaps only an expression of a sound humility in front of users and their professional expertise.

Even with a healthy mixture of deduction and induction, of construction and evolution, serious questions can still be raised. So far we have assumed that the process of systems development begins with a given, well-defined problem. But often we have to begin while still not really knowing what the problem is. Sometimes we cannot really say what the problem is until we have found a solution, and even then we may have our doubts. In practice the process of problem solving often becomes entwined with the process of problem identification.

Once we begin to look more closely at the process of problem identification, we shall find that there are a number of important aspects of systems development that are not even noticed in either the construction approach or the evolution approach.

6

Intervention

Systems developers want to develop high quality systems. Let us take that ambition for granted. This is not meant to deny that our professional activity has other aims that sometimes may come into conflict with this striving for quality. We want to make a profit, and we want our systems to increase the profit for our customers. We are lazy and don't want to work ourselves to death. We want our systems to serve humanity, to make life better for people rather than making it boring and monotonous, to empower and extend the abilities of people rather than diminishing or killing them.

It would be wonderful, of course, if all these aims – profit, our own comfort and career, liberation, democracy, peace, etc. – were to come together in one fundamental aim: quality. But that is not the way the world is. Conflict rather than harmony, contradictions rather than consensus, is what we have to expect when developing computer-based information systems.

If we believe in progress and change, we do not mourn the lack of harmony and consensus. Conflicts and contradictions are instigations to change. It makes the world a tougher place, but it also makes it more exciting. Who would really want to be in heaven? After a few days of listening to the harps, we would begin to fret: "Doesn't anything ever happen here? Where is the action?"

Construction and evolution, as paradigms for systems development, use problems and their solutions to organize the development process.

In systems construction the process begins with a given problem and ends with the realization of a solution. The process is structured, the labor is divided, and general methods are used to determine the structure of the problem and the structure of the forthcoming computer system. The closer the process gets to the solution, the clearer the structure of the computer system becomes, and the more stable the project organization will be. After an initial scurry of creative confusion, the project group settles down.

In systems evolution, the problem is still given but questions are raised. The problem is not formulated and understood in such a way as to have clear-cut criteria for success. Part of the development effort will be devoted to interpreting and reformulating the problem. The process is structured in an open and dynamic fashion in order to support learning. The resulting system is not conceived from the outset. It emerges as a result of concrete experiments with different attempts at partial solutions.

Systems construction and systems evolution are two well-known and complementary problem-solving strategies. In both paradigms, the computer is used to solve a given problem. Installing the computer system is viewed as an instrumental operation upon the user organization. The third paradigm for systems development, which we shall now describe, differs from systems construction and systems evolution in both respects.

From Situations to Problems

Think about the following situation: We have been engaged by a large municipal hospital to start a new project as part of an ongoing organizational change effort. In the project we are expected to cooperate with the unit manager and six nursing supervisors in a surgical unit.

The objective of our intervention is twofold. Our primary objective is to design a computer-based management information system. The system is meant to serve as a personal tool for each of the managers to use in drawing up and maintaining work plans and handling personnel status information for which they are responsible. The system should also serve as a medium for coordination and communication between the seven managers in the surgical unit. Our secondary objective is to develop and implement the new system in the organization, changing work routines and patterns of cooperation between the seven nursing managers.

A couple of years ago the surgical functions of the hospital were reorganized in response to economic and political demands. Previously, this specific surgical sector had been organized as six independent departments, each with its own nursing supervisor. In the reorganization, the departments lost their independence and were turned into six subordinate sections of one centralized unit. The new organization was to be jointly managed by a new unit manager and the six nursing supervisors. The intention was to reduce expenses by sharing resources, and there was also some talk about a shared, central pool of nurses.

This reorganization of the surgical unit was not well planned and implemented. Today, the surgical unit is struggling with the consequences. The unsatisfactory outcome of the reorganization can be blamed on the fact that the nursing staff was not engaged in the reorganization effort. Participation has remained low, and skepticism about the new management principles is still high. In addition, there are substantial differences between the sections. Their tasks vary from 20 percent to 80 percent of immediate surgery, and their planning horizon and management procedures vary accordingly.

In the nursing case, as we shall call it, the challenge is to intervene in this problematic situation, facing the established management traditions, coming up with new approaches, and generating useful computer applications supporting a new management information system. To do this effectively, we have to consider how to design a systems development project together with the seven nursing managers.

Computer systems are going to play an important role in the new management system. But there is no definite data processing problem given, and the key issue in the problematic situation is not a matter of data processing. It is a much more fundamental matter of attitudes, management practices, and concrete forms of cooperation between the seven managers. A new computer system is needed, but to be useful it has to be designed as an integral part of a wider system.

How can we approach the nursing case? The first major challenge is to create a vision of management in the newly designed organizational structure, a vision that is effective in implementing the intentions behind the reorganization and is accepted and shared by the seven nursing managers. How do we create such a vision?

A second more specific challenge is to identify a strategy for using computers as part of this vision. What information is needed in the new management information system? How can computers be used to store and process the relevant data?

Meeting these challenges we have to remember that the beliefs, preferences, and traditions of the seven nursing managers are vital elements in the development effort. The perceptions and practices of the seven nurses have to be appreciated and respected, not only by us but by the nurses themselves.

A highly cooperative effort, in which the nurses are actively involved in analyzing the situation and designing the system, might be the key needed to make change possible. In this process, we can establish cooperative design seminars, we can apply methods like future workshops, or we can use metaphors to stimulate learning and creativity among the nurses and ourselves.

Breakthrough by Breakdown

In the nursing case we are called in to assist in a change effort that has failed. Change is now even more difficult than usual. To open up the situation and prepare for change, we can introduce *metaphors* into the discussion. Metaphors can be used to see well-known situations in new and useful ways. When the nurses look at their management situation as if it were something else, it becomes possible to draw on their knowledge and experience from outside work in exploring possible avenues of change.

The new management information system might be viewed as a production control system for the effective use of local and centralized resources in the surgical unit, or as a booking system to be used by local units to book centralized resources as needed. It could also be viewed as a reporting system to assist the unit manager in reporting to the hospital about the activities of the unit, or as a communication system to facilitate communication and interaction between the seven nursing managers. But it could also be viewed as a manipulation system to support each of the seven managers in exercising power and manipulating colleagues to obtain local advantages and benefits.

For a metaphor to be useful it must, of course, be relevant and rich. There have to be some important similarities between what we try to learn about and the phenomenon we compare it to. And we have to know that phenomenon, and preferably have many concrete experiences with it.

The suggested metaphors are useful, not because they are accurate descriptions of the new system but because they can open our eyes to disregarded aspects of such systems and make us think along new lines. Metaphors make us creative. They are a way of drawing on our experiences from different areas of reality, making fruitful combinations of ideas that we have a tendency otherwise to keep separate.

To deal with a complex environment, organisms rely on a strategy that pays attention only to what is new or different and takes the rest for granted. Human beings are no exception to this rule. The all too familiar becomes practically invisible. We don't really see it until it is gone. It is easier to study a foreign culture, a foreign organization, than your own. It is easier to see the mote than the beam. To study something, you have to view it from a distance, to see it from the outside.

The philosopher Martin Heidegger has used the terms *ready-at-hand* (*zuhanden*) and *present-at-hand* (*vorhanden*) to distinguish between tools that are unobtrusive extensions of our body, and objects that we attend to. Something that is ready-at-hand can become present-at-hand. This happens by itself whenever a tool breaks down. But we can also make it happen by a conscious effort.

The breakdown of a computer system suddenly makes us attend to the system. A little while ago we were using the system, concentrating on our primary task. Now, all of a sudden, the system intervenes between us and the task. Our immediate concern is, of course, to avoid breakdowns. We want to design computer systems that are as robust as possible given our resources. We also want to design computer systems that are easy to learn, so that we may move swiftly from a state of present-at-hand to ready-at-hand. We want the system to disappear as quickly as possible from the view of the user.

But there is another side to this. As computer systems experts we keep the systems we develop present-at-hand. Designing systems to be ready-at-hand for the users means shutting the user out from the design process. When a breakdown occurs, the user is helpless. The more user friendly the technology, the smaller the chance that the user will deal competently with breakdowns.

If a malfunctioning system cannot be easily fixed, the breakdown may turn into an occasion for developing the system, for changing it. By a seemingly perverse method of producing breakdowns, we can create occasions for breakthroughs into new areas of development. And that is, of course, what we do when we experiment with and test new technologies.

By pushing the prototype to break down, we can learn about its weak points and use that knowledge to improve the system. We don't want to use that method in dealing with organizations, but we can take advantage of this very process when it occurs without our doing. And we can be observant when breakdowns occur in organizations, particularly in their use of technology, and use these breakdowns as opportunities for breakthroughs. In addition, there are other methods we can use to make things present-at-hand, to simulate breakdowns by making what is familiar seem alien.

In the nursing case, we can use metaphors in the design process to create breakthrough by simulated breakdown. Metaphors help us make explicit the established traditions and norms governing present management procedures. The existing management information system is no longer ready-at-hand – that is, implicit, functioning in the background as an integral part of being a manager in the unit. The use of metaphors forces nurses to make the management information system present-at-hand – that is, explicit, as the immediate object of their attention.

This process of turning something ready-at-hand into something present-at-hand is normally caused by the breakdown of a tool or instrument, by a crisis in an organization, etc. The use of metaphors is a conscious and artificial way to bring on such a change from being implicitly taken for granted to being explicitly attended to. The result may very well be a breakdown or crisis when what has been made explicit becomes possible to question. Playing with metaphors is a way to stimulate creativity, but the game we then engage in might be a dangerous one.

The metaphors invite the nursing managers to see their work in new ways. Nursing management is neither production control nor a question of booking or manipulation. But the metaphors make the nursing managers see their own work as if it were, and this helps them to evaluate established traditions and to create and consider possible changes in their management organization. As new visions about management of the unit emerge, more specific areas for application of computers are identified.

It is our job to look for *structured domains* of information usage, where stable forms of information or stable procedures for processing information exist or can be established. The identification of relevant structured domains requires technical competence, but the activity is carried out as an integral part of evaluating different proposals for a new management system. The new computer system is viewed as one important part of a wider organizational system.

Change

In the nursing case the problem is vaguely stated, and the computer is not thought of as a self-sufficient instrument. The objective is to increase management competence and the performance of the seven nursing managers. Systems development is seen as a complex intervention into the affairs of the hospital rather than as a simple instrumental operation. As systems developers, we are more like general practitioners, getting involved with the family and working life of their patients, than like surgeons attacking separate organs with a knife.

Systems development turns into organizational action because it is no longer possible for the involved actors to agree even on a very general formulation of the problem. Systems developers are called in because of a general need to improve the organization's information processing and because computer technology is conceived as the standard solution of our time. They are invited to an organizational game, and rather than going ahead to construct a computer system, they may do well to stop and wonder what their role in that game is supposed to be.

The challenge arises not only because of the complexity and uncertainty of data processing problems but because of the complexity and uncertainty involved in organizational change. Different actors within the organization have different, often conflicting, interests, and they give different interpretations of events and different proposals for change. They engage in complex power struggles and play organizational games that are difficult for an outsider to identify and understand.

In an intervention process it is difficult to isolate and make explicit data processing problems, partly because of the holistic nature of organizational situations and partly because of the powerful dynamics involved. Problem owners and users learn as the systems development process evolves, perceptions and preferences change, and the organization itself is in turbulence in a turbulent environment.

Systems developers become consultants and change agents. They are called upon because of their technical competence, but they have to be equally skilled at handling organizational change. They must negotiate and create commitments with other involved actors. Systems development projects are parts of wider organizational efforts, and the resources and contractual arrangements related to a project have to be constantly nursed and defended. The challenge is not merely to cope with bounded rationality. The systems developers must take part in organizational games with mixtures of cooperative and opportunistic behavior. As consultants in the nursing case, we have to realize that the six supervisors have different and conflicting views on planning and management because of the considerable variations in immediate surgery. Also, the new unit is organized in such a way that the sharing of resources can lead to rivalry and conflicts between sections.

The systems developer is no longer an expert solving the problems of other people. The problem owners and users are themselves active and responsible participants in the process. The users have become designers, and the task of the systems developer is to facilitate learning and give technical advice.

Systems construction and systems evolution are approaches to systems development for users. Intervention is an approach to systems development with and by the users. Responsibilities are negotiated and shared between systems developers and users.

The intervention approach stresses quality in use. Quality is not only related to the computer system as such but more broadly to the way the computer system is implemented and adapted in the actual use situation. As a consequence, quality can be improved either by changing the computer system or by changing the organizational context in which the system is used.

Education, training, and explicit attention to attitudes, needs, and preferences are essential elements in successful development projects.

The result of an intervention process is not a computer system representing the solution to a specified problem. A systems development project results in a changed organization. The best we can say is that a new situation has emerged. The best we can hope for is that more useful and effective procedures and tools for information processing have been introduced. The situation is different, and gradually new problems and conflicts will appear.

If we were to characterize the world of the interventionist using only one word, contradiction would be a good candidate. There are contradictions within the project group itself, due to project uncertainty, unclear project aims, different interests in the group, lack of experience in dealing with difficult organizational problems, and so on. There are contradictions between the aims of the project and the available resources, including the competence of the project group, particularly as the aims undergo changes. There are contradictions between the project group and the users and between different user groups in relation to the project aims.

Often these contradictions are treated as nuisances that disturb and threaten the systems development process. But the interventionist takes these contradictions seriously, treating them as opportunities for intervention rather than as nuisances to be neutralized.

Such a dialectic approach to systems development is deeply romantic, viewing the organization as a living organism with powerful internal forces and conflicts. This approach is very different from the mechanistic, hard systems construction paradigm that treats the organization as a machine to be modeled by a computer system. It is also different from the soft systems, evolutionary approach with its view of development as cooperation with the users in a harmonious process of mutual learning.

While a hard systems approach will explain failure by referring to lack of resources, bad technology, and ignorant users, the soft systems approach will blame failure on mutual misunderstanding. They are both convinced that success will be inevitable, given, in one case, more money, technical development, and more education or, in the other case, more time for communication and mutual learning.

Those following a dialectic approach will be less optimistic. Or better, they will have a different conception of success. Unlike the advocate of the evolution approach, they are not really interested in erring less and less and less, since this only means that they are no longer learning. And viewing systems development as a political process, it does not really make sense to speak of failures. One person's loss is another's gain. It all depends on which side you are on. And what is won today may be lost again tomorrow.

The idea of systems construction restricts the responsibility of the systems developer to meeting the requirements explicitly formulated by the client and agreed to in the project contract. Failure to meet the contract can normally be blamed on unrealistic expectations or unsatisfactory technology. The systems developer can be trusted not to promise the impossible and to have a firm control over the technology. The responsibility of this systems developer stops when the system is up and running. What really happens in the organization, the gain or loss in productivity, the way the system is used, is clearly beyond the area of competence of this systems developer.

This is obviously unsatisfactory for systems developers who want to stay in practice, getting new jobs from old clients and new clients on recommendations from satisfied clients. The only too obvious discrepancies between ideas behind the design of computer systems and their actual use, between expectations and reality, well documented for systems installed in the 1970s, eventually led systems developers to adopt a more evolutionary attitude. But the emigration from the safe mathematical, or engineering, world of systems construction to the uncertainty of organizational problem solving has been hampered by the enormous demand on computer systems and the traditional education of systems developers as programmers. As clients become more demanding and professional pride makes itself felt, we are seeing this emigration increase.

When systems developers go one step further and become interventionists, they take responsibility not only for the design of the computer system but for its actual use in the organization. They are not satisfied with meeting the requirements of a contract but want to deliver a system solving the real problems of the client. They see the computer system as playing a vital role in the organization. Their design really makes a difference to the organization, and they have to make sure that they develop a useful system, even if clients may not know at the outset what their real needs are.

As professionals, systems developers realize that in designing and installing a computer system they become responsible for profound changes in the organization. Designing the computer system is really a means of redesigning the organization. Accepting that responsibility, they simply have to make sure that the system has the desirable impact they and their clients intended it to have.

Making It Happen

The construction and evolution approaches are complementary, and according to the principle of limited reduction they should be combined to effectively cope with complexity and uncertainty. The intervention approach suggests a less instrumental view on systems development. The professional challenge is not merely to choose the right combination of approaches. The challenge is to understand and change established traditions in the user organization as well as in the project group and in the development organization as a whole.

The issue of implementation is crucial in the nursing case. The nursing managers have already experienced one failure. The present situation with a formally established, nonfunctioning organization is unbearable. Together with the nursing managers, we therefore have to go beyond a narrow development perspective and make sure that we succeed in establishing new and better management practices in the surgical unit.

The challenge of the nursing case is to design and implement a new management information system that is both operational within the new unit and at the same time acceptable in relation to some of the established norms and traditions. Which of the established management traditions should or must be preserved? And which are to be transcended?

Methods play a minor role in the intervention approach as compared to the practices and norms of the systems developers and users involved. There is an important difference between methods and practices, between what we are supposed to do, what we say and think we should do, and what we actually do. Interventionists are aware of the difficulty involved in bridging these worlds, and they accept this challenge using contradictions to their own advantage in struggling for project success.

In the nursing case, we can distinguish three aspects of organizational practice. The first is the official and *formal aspect* of methods and procedures. We can learn about this aspect by looking into the handbooks and standards of the surgical unit. In our case we should analyze the procedures in use for planning the activities in the sections and for reporting status information to the central hospital administration. And we should design new procedures for making such plans, for reporting status information, and for coordinating section plans with the overall plan for the surgical unit. We should mainly focus on these procedures for management and coordination and for registration and exchange of information. But we also need to look into the very procedures of nursing.

The *opinions* of the nurses are the second aspect of their organizational practice. Here we are concerned with what the nurses think they do or should do. We can learn about this aspect by discussing, with the nurses, both previous, present, and future practices. We need to discuss the differences between the six sections of the unit. What do the nurses think about these differences? To what extent and why are they essential? To what extent are they accidental and therefore easier to change?

Finally, we can consider the *actual behavior* of the seven nursing managers and the deeply rooted assumptions governing their behavior. We can learn something about this third aspect of organizational practice by observing the managers in their daily work. When do they meet? How are these meetings organized? What kinds of informal exchanges take place between them? What are the role of these exchanges and how do they relate to the more formal exchange of information? To what extent are formal procedures followed? What kinds of alternative routines have been established?

The basic assumptions are often taken for granted and are rarely expressed, and the nursing managers, like other organizational actors, tend to be unaware of them. But they still dominate the daily practice within the surgical unit of the hospital. Appreciating these deep-rooted aspects of organizational practice should make us humble as systems developers. In contrast to formal structures and personal opinions, the assumptions and actual practice of an organization cannot easily be changed through an explicit design effort. Fundamental beliefs and practice are expressions of the organization's history, and they develop and change through collective trial and error.

Changing what people believe is always a slow process since it means changing people. And engaging people in serious discussions about what they believe is actually a vital element in such a change process. This way of working, by engaging in a dialogue with the nursing managers rather than in a controlled change process, has its foundation in a romantic belief that changing a person's work habits means changing that person's identity.

In most cases we adopt a more mechanistic way of working, preferring to work with the formal organizational structure, teaching methods and changing opinions, rather than engaging in dialogue, negotiation, and understanding. It takes a lot of courage to attempt a romantic way of working with an organization. It takes courage to engage in dialogue, to really listen. It rarely happens.

Important aspects of our own organizations are so deeply rooted in us that to become aware of them and appreciate their importance we must look at other organizations – or we must create breakdowns in our own organizations. A traditional investigation of the surgical unit will attend only to formal procedures and opinions. To change the practice of the nursing managers, we need to disturb the established routines and involve the nurses actively in developing and adopting new professional standards.

In many situations, individuals and organizations are painfully aware of limitations, inconsistencies, and failures in their behavior. They might even know about alternatives that would change and improve present practices. Still, their actual behavior does not change. Organizational problem solvers explain such failures by referring to limited resources or limited communication abilities. As interventionists, we have a third kind of explanation: In some situations we have the time and knowledge needed to effectively improve our behavior, but our limited capacity for learning makes it difficult to actually change. In fact, a review of the previous reorganization of the surgical unit reveals that this effort seems to have failed due to the limited learning taking place in the organization.

We have used Linnaeus and Darwin to illustrate the construction and evolution approach respectively. Linnaeus, the great systems builder, gave us a taxonomy, a foolproof method for classifying the plants. In his conception of the world as fundamentally unchanging, in his belief in logic and systematic organization, and in his belief in a complete and final method for classifying nature, Linnaeus is an outstanding example of construction thinking.

Darwin is, if not the father of evolution, the greatest source of ideas about evolution. In his conception of the world as constantly changing, little by little, in his belief in induction, in careful observation and collection of facts, and in his very idea of selection as the mechanism of evolution, Darwin is equally outstanding as an example of evolutionary thinking.

It is much more difficult to find a good representative of interventionist thinking. No biologist comes to mind. So we shall have to make do with Hegel, and we are then thinking not of his natural philosophy but of his conception of history as a process driven by a dialectic interplay of contradictions. When as interventionists we approach an organization, we are interested in changing the organization. And we are interested in real change, not just the installment of a new formal structure that will remain a superficial decoration.

In order to really change the organization we have to convince its members that we can offer a way for them to realize their purpose at work. That is, if they are dissatisfied with the way things are, we have to show them a way toward improvement. If they are satisfied with the current situation, or say they are satisfied with it, our task is more difficult. We shall then have to change their standards in order to convince them of the value of participating in a process of change. As we intervene in the organization, in both cases, we can use Hegel's ideas about the dialectic contradictions between what something is and what it is meant to be, the way such contradictions manifest themselves in a social organization, and the way we can play on such contradictions.

Computers Are Our Business

Some would argue that the intervention paradigm has taken us too far away from the technical world of computers and into the social world of organizations. Many systems developers work with purely technical problems, designing, implementing and maintaining computer programs, they would say.

So far we have concentrated on the nursing case, thereby emphasizing intervention in the user organization. But the intervention paradigm applies equally well to project situations and project management. It is a powerful approach to become aware of our own situation as programmers or systems developers, and to change the ways in which our own work is organized and performed.

Let us illustrate this by looking at a situation that, sooner or later, we all have to face. A project is governed by a fixed contractual arrangement requiring certain specifications to be met using a settled amount of resources. Now, due to the many risks and uncertainties involved in a project, it often turns out that more resources are needed to develop the system. How should we act when the need for extra resources has become obvious? Should we keep our own management and the client informed about delays? Should we ask for more resources as soon as the need has been identified?

The intervention paradigm suggests that we participate actively in various organizational games. In these games actors might withhold information to gain a bargaining advantage. And they will often play their own cards more or less opportunistically to defend their own interests. Maybe we shouldn't ask for more resources as soon as the need has been identified?

Considering the psychology involved, we might argue that it is not as much the amount of extra resources needed in our project as the number of times we require new resources that determines the reaction of our management and client. If this is so, we should definitely withhold information and accumulate a solid argument for requiring a considerable amount of extra resources. In the long run this will benefit the project more than would a more interactive approach to communication and negotiation with managers and clients. Some would argue that this is manipulation. Others would argue that it is equally in the interest of managers and clients.

Similar considerations apply to the relation to our colleagues within the project. If you are assigned responsibility as project manager, should you then use a hidden as well as an official agenda? Should you keep hidden a pool of extra resources for critical situations? Different viewpoints are possible and arguments will differ, but the intervention approach at least directs our attention to these problematic situations.

There are, of course, deep-seated differences between designing a computer system and changing an organization. One is the difference between designing a stable structure and creating a dynamic process, between designing a house and cultivating a garden, between writing a textbook and teaching a class, between delivering a product and intervening in a process. And as the difference between these two tasks becomes clear, the systems developer begins to realize that the latter is not really a design task after all.

A social organization is not determined by its use of computers or more generally by its technology. A social organization is a process as well as a structure. To solve organizational problems or, better, to intervene professionally in an organizational process – be it in relation to the user organization or in relation to our own development organization – we need other tools to complement our technology. And we have to use our tools, technical and otherwise, as tools for cultivating a process rather than designing a product.

At this juncture, systems developers are threatened by an identity crisis. Appreciating the limits of design, doubting the omnipotence of computer technology, they are faced with tasks for which they are ill prepared. Educated in the expert construction of computer systems, they are now expected to act as professional organizational agents. Mastering the design of computer systems, they are expected to master the cultivation of organizational processes.

Moving from construction to evolution to intervention is a dialectic process exemplifying the Hegelian schema of thesis, antithesis, and synthesis. The first step in this process involves seeing that the constructed systems seldom achieve what they were intended to achieve. They were designed by experts working in isolation from the people whose working conditions the systems were intended to change. The systems construction paradigm thus exemplifies an authoritarian, uninformed operation on the lives of the users to be.

The move to the evolution paradigm is a move toward designing better systems, but it is also a move toward democratic values. As systems developers then move on to the intervention paradigm, becoming cultivators of organizations, they seem to return to a more authoritarian position. For what is cultivation if not authoritarian?

If evolution is a process of communication and mutual learning between equals, then the cultivation of an organization is the conscious manipulation of people's ambitions and desires in the interests of the organization – a bit like horse breeding it would seem. As organizational agents, systems developers use the authority they have as computer experts to further the interests of the organization, often against the immediate desires of its members. In this respect they resemble the teacher using her authority to make her students study when they would rather play, thus furthering what she believes to be their long-range interests.

We can use this analogy to see how to solve the dilemma of rejecting one authoritarian position only later to end up in another. The process taking the systems developer from construction to intervention is a dialectic process, meaning that previous positions are not really rejected but retained in, hopefully, a healthy mixture. As an interventionist, the systems developer remains a problem solver relying on both construction and evolution approaches.

We have described three different paradigms for systems development, indicating their strengths and weaknesses, pointing out their presuppositions about reality and their conceptions of how systems development should be done. As a way of introducing

de·vel·op (dǐ věl′əp), *v.t.* **1.** to bring out the capabilities or possibilities of; bring to a more advanced or effective state; cause to grow or expand; elaborate. **2.** to make known; disclose; reveal. **3.** to bring into being or activity; generate; evolve. **4.** *Biol.* to cause to go through the process of natural evolution from a previous and lower stage, or from an embryonic state to a later and more complex or perfect one. **5.** *Math.* to express in an extended form, as in a series. **6.** *Music.* to unfold, by various technical means, the inherent possibilities of (a theme). **7.** *Photog.* **a.** to render visible (the latent image) in the exposed sensitized film of a photographic plate, etc. **b.** to treat (a photographic plate, etc.) with chemical agents so as to bring out the latent image.—*v.i.* **8.** to grow into a more mature or advanced state; advance; expand: *he is developing into a good citizen.* **9.** to come gradually into existence or operation; be evolved. **10.** to come out or be disclosed: *the plot of a novel develops.* **11.** *Biol.* to undergo differentiation in ontogeny or progress in phylogeny. **12.** to undergo developing, as a photographic plate. Also, **de·vel′ope.** [t. F: m.s. *divelopper.* f. *di-* DIS¹- + m. *voluper* wrap. Cf. ENVELOP] —**de·vel′op·a·ble,** *adj.*

the paradigms, we have discussed three different paradigmatic examples: the 8-Queens problem, the Cocomo problem, and the nursing case. These examples help us remember some of the fundamental differences between the paradigms.

The examples are fundamentally different, but what is different is the approach taken, not the problem situation itself. There are no clear criteria for when to use the different approaches. Paradigms are not the kinds of things you switch back and forth between. They are not instruments to be selected in response to the type of situation we are confronted with. People become heart surgeons or general practitioners; they do not switch back and forth between the two as the situation changes.

Each paradigm demands a specialization that is hard won and needs constant cultivation. With this specialization comes also an identity and a particular view of the world that is not something to be changed willy-nilly. This inertia is a fundamental idea behind the notion of a paradigm.

When we think of a paradigm as an instrument to be chosen, we express the position of the first two paradigms for systems development with regard to computer technology. Treating our major asset, the computer, as an instrument, we tend to generalize our instrumentalism. When we argue that a paradigm is more like an identity, a style, a world view, we illustrate the position of the third paradigm for systems development with its view of the systems developer as involved in a complex social interaction.

If you cannot choose between these paradigms, what is the use of going on and on about them, as we have done in this chapter? Instead of answering that question, let's take a closer look at how these paradigms relate to one another. One way of doing this is by reviewing them in terms of the three ways of systems thinking we have discussed: the hard systems approach, the soft systems approach, and the dialectical systems approach.

It is not obvious how the three ways of thinking should be applied to the three paradigms. Let us try a static alternative first. Systems construction is obviously a good example of the mechanistic world view and the hard systems approach. Intervention is clearly a good example of the romantic world view and the dialectical systems approach. But what about systems evolution? We can think of it as some sort of compromise between the other two, conceived as extremes. On a foundation of mechanism and hard systems thinking, this paradigm introduces soft systems thinking, but for mainly pragmatic reasons.

With this classification, the evolution paradigm becomes the central paradigm. Our task as systems developers is problem solving in the complex context of organizational information processing. Sometimes the problem is extremely clear cut, and we can adopt the strategy of top-down systems construction. Sometimes the situation is so uncertain that we have to practice intervention. But most of the time we are in between these extremes doing experimental problem solving.

This approach to the three paradigms also invites us to think of them as stages on the way toward a functioning computer system. Ordering them from left to right, we want to end up on the far left with a routine computer construction task. The less well-structured and uncertain the situation is, the further out to the right we will have to begin, and the longer it will take to get back to the business of constructing computer systems.

Another way of thinking of the three paradigms, but still within this static view of extremes and compromises, is in terms of the different disciplinary origins of systems developers. When computer technology, in the 1960s, began to be used for the design of information systems, computer engineers moved out into organizations and business school graduates began to study computer technology. With one group beginning at the systems construction end and the other at the intervention end, they might eventually meet in a pragmatic position based on the evolution paradigm. But this is a misunderstanding, either of what the intervention approach is all about, or of what is normally learned at a business school. Graduating from such a school, students will often go out into the world with hard systems ideals – with a belief in formal organizations, methods, and procedures. These ideals are as strong as those of engineers, even if based on a much weaker and more flimsy basis.

A more dynamic way of thinking views the three paradigms as stages in a process of development. We can think of them as stages in the history of the practice of systems development, or we can see them as stages in the career of computer engineering students broadening their competence to include information systems science. Less concretely, we can view the three paradigms as stages in a dialectic development of ideas.

Beginning with a belief in computers and a rather restricted view of what systems development is, we become aware of the role of human beings in putting the technology to good use. We realize that only by working closely with the users will we be able to develop useful information systems. We turn from systems construction to systems evolution.

As we bring in the users, they begin to eat us up, and eventually we have our hands so full of organizational problems that we have no time to deal with computers. Worse, we begin to lose faith in technological solutions. We have now reached a true antithesis to our very first position. We have given up everything that we believed in when we started out as systems constructors. We have changed from a mechanistic belief in hard systems thinking to a romantic belief in dialectic thinking.

But having thus become dialecticians, we cannot settle down. As good dialecticians, we have to move on, and the only way to go is back. That is, we have to surpass our thesis-antithesis with a grand synthesis. We have to find a way of rethinking computers in terms of our hard-won experience, to obtain a deeper, constructive understanding of the role of this technology in changing organizations.

Part III

Quality

The fundamental idea of systems development is to use technology to make the world a better place for us. Systems development is part of the project for progress. And since computer technology is the most dynamic and expansive technology today, software development is currently an important means for progress.

We have begun to use the word quality as a catchall term for good things: quality of work, quality of life, quality system. So, progress is the increase of quality. But what is quality?

Asking, in part III, what quality is and how it is achieved, we begin in chapter 7, by looking rather closely at the products of software development and the quality methods we use. We discuss the quality of technical artifacts more generally and distinguish between functional, aesthetic, and symbolic quality, and we address the difficult issue of the objectivity of quality. We argue that quality has to be thought of as a challenge, something to be strived for but never reached.

In chapter 8 we turn to the organizational context of the development and use of computer artifacts. We use as our example an organization producing computer systems, asking how quality can be achieved by improving the professional standards of software development organizations. We discuss what it takes to change the culture of an organization, be it a software development organization or a user or client organization, and how our striving for quality will involve us in a difficult struggle with customers, colleagues, technology, our own ideals and habits, and the idea of quality itself.

By shifting our perspective from the artifact itself to its cultural context, the discussion of quality becomes more realistic. When, in chapter 9, we look even further and ask more generally about the use of computers in organizations and society, questions of quality become even more complex. These questions lead us into difficult moral and political issues concerning when and how to use computers and the role of this technology in changing society and the lives of all people. Professional systems developers want to do a good job, but according to what standards?

7

Artifacts

All users of computer artifacts are painfully aware of the low quality of software. There you are putting some finishing touches to your paper before printing it, and since you have the time you decide to take a break and install the new version of the operating system on your personal computer. At the beginning, things go smoothly, but all of a sudden you are caught in what seems to be an eternally branching questionnaire, having to answer endless questions about things that seem wholly irrelevant, and that you know nothing about.

All you wanted to do was very quickly update the system – for no particular reason – and now you find yourself on a slow march through the whole system. Something was said in the very beginning about how you could exit the process by giving a specific command, something you did not pay enough attention to, or did not bother to write down or try to memorize. Now, you try out all the commands you can think of that normally will terminate a process, but none has the desired effect.

Half an hour later you finally get back to real work, only to find that under the new system you can no longer use your thesaurus. You begin to regret having updated the system in the first place. Suddenly, you make a worse, indeed catastrophic, discovery: You can no longer print out documents from your text processing system. Sweat is breaking out all over. Now, how about that paper?

You try to calm down, and begin to think about reinstalling the old system. But how do you do that? Do you still have those old disks lying around somewhere? How long will it take to reinstall the old system, and can you do that without ruining what you have on the hard disk? You feel like a fool getting caught in a trap like this, and not for the first time. All you wanted to do was write a paper, a paper that had to be finished today.

Systems developers have, of course, a professional concern for quality, and several attempts have been made to describe and catalogue the different aspects of software quality. Quality is then often analyzed into a number of factors, such as correctness, reliability, efficiency, integrity, usability, maintainability, testability, flexibility, portability, reusability, and interoperability. The first five factors relate to using the software; the next three factors address modification of the software according to new requirements; and the last three concern transformation of the software to operate in new environments.

Each factor is explicitly defined in order to give a precise meaning to the terms used. Correctness reflects the extent to which a program satisfies its specifications and fulfills the user's objectives. Maintainability represents the effort required to locate and fix an error in a program. And portability is an estimate of the effort required to transfer a program from one hardware configuration to another.

Some of these factors contradict each other, which means that one of the practical difficulties in developing quality software is to decide on the trade-offs between the relevant factors. The more you strive for efficiency, the more difficult will it be to obtain satisfactory maintainability and flexibility.

The quality of one and the same system is normally judged differently by the actors involved in its production and use. Different weightings of the quality factors will result in different evaluations of the system as a whole. In addition, the systems developer may be satisfied with the quality of the produced system while the customer is dissatisfied with its quality in actual use. The users may be satisfied with the quality of the development process viewed as a learning experience while dissatisfied with the actual system as it is put to use.

The systems developer can hide behind the specification, claiming that the customers have no grounds for complaint since they are getting what they asked for, while the customer can complain that the specification did not have a good enough quality. Discussions like these are frustrating, and the easy way out is simply to define quality as meeting the expectations of the customer. But to choose that strategy to deal with quality will not gain you the respect of your profession.

The idea of professional expertise involves the competence to evaluate the quality of a system irrespective of what the customers have to say. While their expectations of the system naturally play an important role in the design, it is the responsibility of the systems developer to adjust those expectations to considerations that go beyond the competence that can be demanded of the customer.

Being responsible for the introduction of a complex and rapidly evolving technology into an organization, the systems developer is responsible for the future consequences of the use of that technology, for problems of maintenance and further development, for compatibility with future technology, and for the usefulness of the technology as the organization changes. This is a major responsibility and some of it can be shared with the customer, but to rid oneself of all of it by a policy of only giving the customers what they say they want is to betray the professional standards.

If the quality of a system cannot simply be evaluated by its success in meeting the expectations of the customer, how then can it be judged? What can we mean when we say that a system is of good or high quality? What is quality and how do we produce it?

Quality Methods

Computers, programs, models, and prototypes are all *artifacts*. They are manufactured objects rather than natural phenomena, and they are made to fulfill a purpose. Systems developers are responsible for the quality of these artifacts. They are responsible for the quality of the computer system as a technical object delivered in response to a more or less formal contract with a client. They are responsible for the quality of the computer system as a useful tool for its users. And they also have to worry about the quality of all sorts of intermediate products like design proposals, plans, and prototypes.

The obvious answer today, in the systems development community, for how to meet this responsibility is quality control. Systems developers now have methods to help them organize their work and ensure that the artifacts they produce have the quality demanded by their professional community.

The general idea is to control the entire development process from a quality perspective by relying on general methods, repeating the same procedure each time an artifact is produced or further developed. We begin each production process by *planning* for quality: The desired qualities of each artifact, together with the criteria and procedures for evaluating it, are specified, as far as possible, and these specifications are organized into baselines describing important intermediate states of the project.

During the actual *production* of the artifacts, we can improve the quality of our work by applying systems development methods and tools. In addition, various kinds of inspections must be performed to provide status information and intermediate evaluations. When the artifact has been produced, it is time for *evaluation* and *correction*. The artifact is evaluated against its specification, and, depending on the outcome of this evaluation, corrective actions are taken.

Each baseline specifies a number of intermediate objects. A design baseline could, for example, contain specifications of a design document, specifications of a number of prototypes, and a test plan. These specifications are given as criteria and procedures for evaluating the quality of these objects. Thus, for example, it could say that the design document must be in accordance with the requirements document and the company standard for documentation, and that it should be useful as a reference manual during implementation. And it could say that the design document is to be evaluated on these criteria in a formal review performed by both client representatives and responsible programmers.

By making the desired outcome of the production process explicit in this way, two important effects are achieved. We force ourselves to plan, in detail, the evaluation of the artifact, thereby supporting systematic evaluation of documents and systems as an important part of the development effort. Equally important, we help ourselves produce better documents and systems by making explicit criteria that we can use during their production to focus our attention and make priorities.

In general, we use two different forms of evaluation, one based on metrics and one based on experienced opinion. Let us look quickly at each of these before describing in more detail some of the more specific techniques involved.

One of the most powerful ideas of quality control is to turn the general question of quality into more operational questions of specific qualities. Evaluation is then performed by measurement, using established and agreed upon measures or metrics to produce the evaluation. When specifying the artifact, its quality is analyzed into different qualities or quality factors, like efficiency, maintainability, and portability, and further into measurable attributes, like consistency, modularity, and execution efficiency. The quality factor efficiency, for example, can be decomposed into execution efficiency and storage efficiency, and even more specifically we could consider whether data are grouped in the program so as to support efficient execution. For each quality factor, a procedure is then provided for how to combine individual measures into one quantitative evaluation.

This kind of measurement is intended to be objective in the sense of being independent of who performs it. Again, while this ideal is obviously worth striving for, it is important to realize how difficult it may be to attain. We try to substitute subjective judgments for objective measurements by turning general questions about quality into specific and detailed questions about measurable attributes, but the measuring of those attributes may still involve subjective judgment. To use the example above, it is not evident that two different programmers always will give the same answer to the question of whether the data in a program are grouped so as to support the most efficient execution. More generally, it remains an open question to what extent it is possible to structurally decompose the quality of a computer system into such measurable attributes without leaving out important aspects of quality.

Our quality control methods also include techniques for evaluating artifacts that are less focused on detailed specification and measurement of qualities. These techniques rely on competence and experience rather than on the use of metrics. A useful way to evaluate artifacts is simply to ask for someone's opinion, thus depending on that person's judgment and experience.

When we judge an artifact, we use our competence and intuition to evaluate it. The result of the evaluation is a subjective opinion on the quality of the artifact. Such an opinion is useful in itself. It can serve to remind the producers of additional issues to be considered or it can serve to identify mistakes or flaws in a design. Because of the subjective nature of a judgment, its value for the producers might increase if the identity of the evaluator is revealed. But sometimes blind evaluations are preferred to encourage evaluators to give honest opinions. We can even argue for double blind evaluations, to prevent the evaluator from evaluating the producer rather than the artifact. Such double blind reviews are normal practice in the evaluation of contributions to scientific conferences and journals.

Other evaluation techniques, such as prototyping and versioning, are based on experience. Instead of judging the artifact, we try it out. The evaluation is then an expression of our experience in using the artifact. The outcome of such an evaluation is dependent on how it is planned and organized, and it is also strongly dependent on the background of the evaluator.

Evaluations based on personal opinions raise a number of different questions. How do we choose the persons to participate in our evaluations? How do we prevent social relations and personal preferences from influencing the evaluation process in a negative way? How do we transform personal experiences and opinions into constructive proposals for improving the artifact?

We have a number of specific quality control techniques for the evaluation of artifacts. The classical techniques are proofs, tests, reviews, and experiments. Proofs and reviews are both formal, rational techniques, whereas tests and experiments are empirical, practice based.

The software engineering community has attempted to apply the mathematical notion of formal *proofs* to computer programs. Given a program and a specification of the program, an evaluation is performed by proving that the program satisfies the specification. Quality is seen as relative to specifications and the intention is to produce an objective evaluation independent of the evaluator.

In a *review,* the document is given to the evaluator together with criteria for its evaluation. The reviewer reads and reflects upon the document to prepare a personal evaluation. Subsequently, a review meeting is held and the evaluation is presented together with other evaluations. The producers of the artifact participate in the meeting, but they are not allowed to engage in discussions with the reviewers. A review report is produced containing a conclusion on whether to accept the document in its present form. In this case the evaluation is subjective, based on the experience and judgment of the individual reviewers. Quality is considered relative to given criteria but the reviewers can include other criteria as well.

Tests are used to evaluate programs. A test plan is developed together with a set of test data, and the evaluation is then based on the performance of the program on these data. Tests are effective in identifying errors and bugs, but a test is always restricted to the specific data and the specific plan applied. We have to be careful about making general statements about a program based on a single test. Programs cannot be tested to ensure that they do not contain bugs or errors. With a test, we evaluate a program or a computer system as such rather than its quality in use. The evaluation is objective and relative to the criteria implicitly given in the test plan.

With *experiments,* computer systems can be evaluated in actual use. The experimental situation can be more or less realistic, and the evaluator can be more or less representative of the future users. An evaluation based on experiments is different from an evaluation based on practical use in that it is planned. In fact, planning is a crucial factor in ensuring quality in experimental evaluations. Experiments are, like reviews, subjective evaluation forms, and quality is considered relative to planned criteria even if the experimenter can include other criteria as well.

When we go from proofs and tests to reviews and experiments, we make our evaluations less formal and more dependent on intuitive judgment. Our evaluations also become more open-ended, less restricted by explicitly formulated criteria for success.

Proofs and tests are the natural ingredients of a construction approach to systems development. The computer system is evaluated in relation to how well it meets the specification. This can be done formally by using mathematical methods applied to the program viewed as a mathematical object, or it can be done by test runs of the implemented program. In both cases we are interested in determining the internal consistency of the computer system (program) and how well it meets the given specification.

Reviews and experiments become interesting when the computer system is to be evaluated in relation to a wider context involving the experience of users and other experts. We use reviews to determine the quality of a computer system as judged by experienced systems developers or users, and we use exper-

"The beauty of the system is that there are a few small errors programmed into it, which helps to avoid total depersonalizations."

iments to determine the quality of a computer system in actual use. In both cases the given specification plays an important role, but the evaluation goes beyond the specification in adding less formal criteria of success relating to our responsibility to construct good systems over and above what may have been originally specified.

Methods for quality control concentrate on the specification, production, and evaluation of artifacts. We have developed a variety of principles and techniques ranging from rational to experimental and from objective to subjective. But all of the techniques share a primary concern for the artifacts themselves, and for their intended or specified function. The methods strive for quality of the documents, prototypes, and systems we produce. In the following chapters we will focus instead on the integration of these artifacts into their social contexts, on the actual use of computer systems and the social, moral, and political issues such a perspective raises. But even when we focus on the description and evaluation of isolated artifacts and consider traditional ideals of quality control, there are difficult questions that we ought to have answers to. Let us look more closely at some of them.

Functional, Aesthetic, and Symbolic

Computer systems are technical artifacts. They are objects made for a purpose, and we are, therefore, interested in deciding whether they fulfill that purpose or not. This is not true, in the same sense, of natural objects. If we ask about their quality, we have given them a purpose and turned them into artifacts.

The Greek philosophers, notably Plato and Aristotle, would disagree with what we just said, believing as they did in a teleological world, in a nature saturated with purpose. The quality of something, they said – be it an artifact, an organism, or an organ – is measured by its success in fulfilling its purpose.

Today, we have no qualms speaking about the purpose or quality of a kidney or a horse, but we try to keep the distinction clear between artifacts and natural objects. The quality of a horse is relative to a certain artificial use of the horse, for steeplechasing for example. The quality of a kidney is relative to its role in an organism viewed as a functional system. Defining the purpose of the organism as survival, we measure the quality of that organism's organs by their contribution to its survival. But we don't believe that the organism really has a purpose.

We can agree with Plato that the quality of a flute player is determined by how well she plays the flute, since a flute player can be viewed as an artifact with a defined purpose. But we cannot go along with Plato when he says that the quality of human beings, their moral goodness, can be similarly defined. Our primary interest in natural objects is in what they are like, what qualities they have. Our primary interest in artifacts is how good they are, what quality they have.

When we evaluate artifacts, the basic question is always the same: "Is it good?" But different types of artifacts are evaluated by different standards. Technical artifacts, artifacts for everyday use in, or outside, working life are, in principle, evaluated by three different standards: functionality, aesthetics, and symbolism. With functional standards, we evaluate the practical use of an artifact; aesthetics is a matter of how it looks; and symbolism has to do with its social use, what it means to us and signals to others. In practice, these standards are treated very differently, and the weights attributed to them vary.

To decide if shoes are good, in the functional sense, we must know their function. If they are for walking, then they should be sturdy and water resistant, and they should have a rubber sole for good grip, and so on, depending on where the walking is to be done. If they are for dancing, they should be light, with a leather sole to glide on, and so on, depending on what kind of dancing they are meant for. When we try on the shoes in the store, we walk back and forth in front of the mirror, trying to determine both fit (functionality) and looks (aesthetics) at the same time. The clerk will assist us: "Those shoes really look good on you, sir." But then again, we may decide to buy the basketball sneakers, after all, even if they really look terrible and are very uncomfortable, and even if basketball is not our game, simply because "everyone else" wears those kinds of shoes (symbolism).

In general, we tend to believe that while functionality dominates working life, in our private lives we are often more interested in aesthetic and symbolic qualities. But is this really true? Buying a lawn mower or a dishwasher, we look for functionality. Architecture is one area in which corporations tend to stress symbolic value at the expense of functionality. And the uniform of the office worker, the suit and tie, is symbolic rather than functional.

Aesthetic qualities are peculiar because of the self-sufficient, disinterested nature of aesthetic experience. Aesthetic objects are there just to be experienced; you cannot really do anything with them. It is against the nature of a corporation to be disinterested. So when a corporation pours money into aesthetic qualities, it is always for an ulterior reason of a symbolic or, eventually, functional nature. The beauty of our building is not for us and others to just savor; it is there to show the world our class, and if they see that we have class, they will want to do business with us.

When we evaluate the quality of a computer system, and particularly when we try to meet the expectations of a customer, it is perhaps all right to concentrate on functionality, but we might do well to cast a glance in the direction of aesthetics and symbolism.

Treating technology as a symbol rather than as a tool means looking at it as a means for communicating the culture of an organization. Heavy investment in a new technology can be motivated by a desire to be seen as a successful, progressive organization, to impress both customers and employees. Technology will acquire symbolic qualities in many different ways, typically deriving from the means of introduction. How is the decision to invest in the technology made? What part of the organization is behind the initiative? What is the public image of the technology?

Similarly, aesthetic qualities seem to play important but subtle roles of which we seem mostly unaware. Designers outside computer technology are normally engaged in decorating rather than in engineering, but in systems design we tend to be oblivious to the value of an aesthetically pleasing or interesting interface. As long as computer artifacts are being produced by engineers for the use of engineers, this lack of interest in aesthetics may have a natural explanation in the engineering attitude. But when computer artifacts enter the consumer market, money will be spent on decorating the artifacts themselves, not only the sales brochures.

To predict the functionality in actual use of a newly designed computer system, it may very well be important to first evaluate its symbolic and aesthetic qualities. That is to say, the symbolic and aesthetic qualities of an artifact may interfere with its functional qualities. An artifact may be too ugly, or it may be symbolically wrong for us to use it to its full capacity. Similarly, the functionality of an artifact will influence its aesthetic and symbolic qualities.

The stewardesses on Scandinavian Airlines will pour your coffee from a rather unimpressive looking pot. The design of the pot is Scandinavian, all right, but you would probably not notice it. Unless, that is, you know the story behind its design, and the amount of thinking that went into designing this extremely well-functioning coffeepot. The pot is broad at the base and narrows toward the top, thus making for a low center of gravity, making sure that there will be no involuntary tipping and accidental pouring of coffee. The spout is long and attached to the bottom of the pot, so that the pot doesn't have to be tipped very much to pour the coffee and it is drip free. All these factors will ensure that the pouring of the coffee is easy and extremely well controlled. Once you have taken all this in, the pot begins to look much more elegant and beautiful.

Shoes and coffeepots are everyday, simple artifacts for which the three dimensions functional, aesthetic and symbolic seem perfect. More complex technical artifacts, like transportation systems, communication systems, and computer systems, have effects, intentional and unintentional, that go far beyond these three dimensions. Technical systems form artificial environments, worlds that have properties such as being benign, comprehensible, friendly, comfortable, dangerous, and so on. Creating such environments is to shape peoples' lives, and this fact gives to the designer of technical systems a political, ethical role.

Our transportation systems are important parts of society and we can avoid them only by emigrating. If you happen not to like automobiles, you can refuse to use them, but they will continue to shape your everyday life nevertheless.

A technical system has to be evaluated not only by its functional, aesthetic, and symbolic qualities but by its *politics,* its political qualities, as well. Computer professionals usually ignore the aesthetic and symbolic qualities of computer systems. When we evaluate a system, we tend to focus on its functional quality, its usefulness. No wonder we remain completely unaware of its political role.

Objective and Subjective Qualities

There is an important difference between describing an artifact and evaluating it. When we describe an artifact, we attribute certain properties or qualities to it. The underlying assumption seems to be that these qualities are really part of the object. Our evaluation of it, however, goes beyond the mere attribution of qualities.

The quality of an artifact is not just another quality possessed by the object in addition to its color, size, shape, and complexity. The aesthetic quality of a painting is not a quality that we can point to as we can point to other qualities. Size, complexity, shape, and color seem more objective, more real, in the sense of being somehow part of the object. The aesthetic and symbolic qualities of an object seem more subjective, more in the eye of the beholder.

Aesthetic evaluations reflect attitudes and personal preferences, we say, rather than the real properties of the artifact. When it comes to judging the size of an object, however, we think it silly to resort to personal preferences, even if we are aware of the important role of perspectives. Beyond all perspectives, the object is said to have a definite size. Not so with the aesthetic qualities, we say. But then we become hesitant. Some of us find it utterly ridiculous to treat the aesthetic quality of, say, the violin concerto by Beethoven or a self-portrait by Rembrandt as a matter of taste. These artifacts simply have quality.

The functionality of an artifact seems more objective than its symbolic and aesthetic qualities. That the functionality is revealed only when the system is used is no reason to call it subjective. This only means that the functionality of an artifact is a dispositional property, but so are many objective properties, like solubility, magnetism, and plasticity. A dispositional property of an object will show itself when we treat the object in a certain way.

Certainly, it should be possible to evaluate the functionality of a technical artifact without resorting to attitudes and personal preferences. Isn't the attribution of functionality a case of description rather than evaluation?

When we look closer at the functional quality of computer systems, this notion explodes, as we have seen, into a complex mess of different qualities. We want computer systems to be correct, reliable, efficient, easy to use, flexible, portable, and so on, and this list grows and undergoes changes as the technology and its use develop. Evaluating the functional quality of a computer system involves a complex weighing of qualities like these. Are they all objective qualities and are the principles used in weighing them objective? Or does the attribution of qualities like these and the weights attributed to them reflect personal preferences only?

Our attempts to answer these questions are complicated by the fact that the objectivity of qualities generally can be questioned, and indeed was questioned already from the time of Plato. "Man is the measure of all things, of the things that are that they are, and of the things that are not that they are not" said the great sophist philosopher Protagoras, meaning by this that it is for us to decide what qualities to attribute to objects. There is no fact of the matter, disagreement is irreconcilable, a rational decision is impossible. All qualities are subjective, relative to the subject making the attribution.

Philosopher-scientists like Galileo and Descartes tried to reconcile the objections raised by Protagoras with their idea of an objective science of nature by distinguishing between different kinds of qualities. Among the qualities we experience in objects are some that are objective, like shape, velocity, and mass, and others that are subjective, like color, warmth, smell, and taste. The shape we experience is also the shape of the object, unless we are mistaken. But the sound we hear is nowhere in the object. And analogously, computer systems have objective qualities like response time and program size as well as subjective qualities like being easy to use and easy to learn.

This distinction between objective and subjective qualities played an important role in establishing Newtonian mechanics as objectively true, by declaring all the qualities used in that theory to be objective. But when the distinction later was questioned, the objectivity of science was undermined. Sound is a subjective property. The world itself is silent. A human being (or a comparable organism) has to be present for there to be sound. But what grounds do we have for saying that shape, for example, is different from sound in this respect, that it is an objective quality? How do we know that the world is shaped? How do we know that time and space are qualities of the world rather than qualities in our experience only?

These are difficult philosophical questions, and we are lucky that we don't have to answer them in order to arrive at a workable conception of the quality of computer systems. We can simply say that the distinction between objective and subjective qualities is a matter of consensus, relative to a certain community.

This may very well be what Protagoras wanted to say, illustrating as he did his thesis by examples drawn from different cultures, different communities. When a community – for example, a project group or representatives from a software house and one of its clients – agrees on methods of measurement and other criteria for the attribution of certain qualities, those qualities become objective. The qualities for which we can give no such criteria, and the attribution of which we cannot defend by formal argument, will be subjective.

A simple example used by Protagoras to show that "man is the measure" can illustrate this idea of making objectivity relative to a community. If you dip your hand into a bucket of cold water and then dip it into a bucket of tepid water, the tepid water will feel hot. But if you first dip your hand into a bucket of hot water and then into the tepid, it will feel cold. Which situation is real? Well, isn't it obvious?

The water is really tepid, but it feels either hot or cold depending on your previous exposure. But then, of course, the reason we say the water is tepid is that it feels that way if, first, you hold your hand in normal room temperature for a while and then dip it into the water. So the tepidness also depends on previous exposure. Defending your view that the water is hot, cold, or tepid, you can refer to a certain procedure that will produce in any normal person an experience of heat, cold, or tepidness. Depending on the customs of that person, on his previous exposures, he will either reach the same verdict as you or he will disagree. All of you having the same previous exposures, having the same background, will form a community for which the water will be objectively hot or cold or tepid.

Let us return to the fundamental question: Is the functional quality of artifacts objective or subjective? This question is, as we have seen, complicated by the observation that the distinction between objective and subjective qualities is not all that clear.

And, in practice, the distinction between objective and subjective is not really that interesting. Both our objective descriptions of computer artifacts and our evaluations of them are possible because we can agree, in the systems development community, and in specific software organizations and projects, on criteria and standards for description and evaluation.

The question of whether or not the quality of a computer system is an objective quality turns into a challenge to *make* quality objective by developing methods of measurement and other ways of evaluating quality that win the acceptance of a community. The important issue is not whether to use mathematical proofs or formal reviews or whether to perform tests or arrange experiments. The real challenge is to make the involved actors agree on which techniques and standards to apply.

We should not strive for objective measures at all costs. The blurring of the distinction between objective and subjective qualities generally implies a blurring of the difference between objective measurement and subjective judgment. Methods of measuring quality are no more objective than the communal, intuitive judgments of experts or users.

Objective qualities are preferable since they make possible discussion, negotiation, improvement, and cooperation. But if we really want to strive for quality, we cannot turn our backs on even the most subjective, idiosyncratic evaluations, to the extent that such evaluations play a role in our attitudes to computer systems.

Striving for quality does not simply mean striving for objective criteria and standards of measuring quality. It does not even mean striving for objective evaluations, in the sense of communal evaluations of quality. It rather means striking a sensible balance between the most objective and the most subjective, of knowing when to strive for agreed upon criteria and evaluation procedures and when to listen to the idiosyncrasies of colleagues and customers. And it means knowing how to create opportunities for the involved actors to communicate, negotiate, and agree on the quality of the computer system they are engaged in developing.

Striving for Quality

The traditional world was a world of purpose, a teleological world. In such a world, things are defined by their unchanging essence. Something is good or bad depending on how well it fulfills or realizes its purpose or essence. This is true of everything there is, including natural objects, artifacts, and people. When a thing changes, it is either a natural movement, or growth, in the direction of realizing its true purpose or an unnatural aberration away from its essence. The ideal situation is a stable world where everything fulfills its true purpose. Change is always a sign of failure, of lack, either in the sense of introducing lack or remedying a previously introduced lack.

How different is not our modern world. Things don't have purpose; they are given purpose by us. The world itself has no essence; the order it has is made by us. This world view is articulated most clearly by existentialists like the French philosopher-writer Jean-Paul Sartre. Contrasting his view of the individual with that of Aristotle's, Sartre describes the human condition as having no essence: "Existence precedes essence."

Human beings do not exist to fulfill a predetermined essence. They have to design their own purpose, always going beyond what they already are. That is to Sartre the true meaning of the project of life. That life is a project means that we live by transcending what we are, throwing ourselves into the future, setting future goals only to replace them with other goals when we reach them.

In a traditional world we act in accordance with tradition, we do what we have always done. In a modern world we put a premium on initiative, on breaking with tradition. Our actions are future-oriented, goal-directed, rather than oriented by the past, habitual. We formulate goals for our actions and strive to reach them, but then again we are not content to reach them. We formulate goals and struggle for them in order to be able to transcend them.

The traditional world was complete. The modern world is forever becoming. There is no end to change and progress. Like the individual in Sartre's existentialism, everything in the modern world lacks essence. Quality is no attribute; it is a challenge to transcend. As soon as we have given a definition, indicated criteria, we begin to hesitate: Is that really quality?

A lot of our knowledge is tacit, unformulated. Our actions are to a great extent based on know-how, rather than on explicitly formulated rules and principles. Our perception of quality and our evaluations express values that we rarely try to bring into the open. And in practice it is often more fruitful to let our tacit knowledge and values remain tacit.

We have made this point by discussing the difference between what we called Platonic and Aristotelian concepts. A lot of our knowledge is based on Platonic conceptions, on exemplary instances or paradigmatic cases, rather than on Aristotelian concepts, explicit rules and definitions.

But if we want to develop our knowledge, to question and change our values, we must confront them by trying to make them explicit. We must try to turn our Platonic intuitions into Aristotelian concepts. This is exactly what we are encouraged to do by our quality control ideals. This is also what is done in ethics, when philosophers try to define the morally good, what it is to be a good person, to act morally.

All such definitions are threatened by the so called "open question argument." Given any definition of the good, it always makes sense to ask: "You say good means X, and here we have an instance of X, but is it good?" Every attempt to define the morally good or, for that matter, to define quality more generally, will break down and reveal a lack to be dealt with by further attempts at definition.

The open question argument can be taken to indicate that good is not an attribute, not a quality possible to define in terms of other qualities, but rather a challenge to be transcended. In other words, our Platonic intuitions can never be fully captured by Aristotelian concepts.

Our perception of quality and our actions to achieve quality are governed by Platonic intuitions. In order to develop those intuitions we must try to turn them into Aristotelian concepts. We must define quality in order to perceive what is lacking, and we must define quality and intend to achieve it in order to realize that we did not succeed.

This idea of definition and transcendence is at the heart of dialectic thinking, as we find it in both Plato and Hegel. Hegel turns it into the simple formula of quantitative change turning into a qualitative leap. Trying to define quality in terms of qualities, we are forced to add new qualities to our definition in order to remedy its shortcomings, in order to handle counter instances. Sooner or later our definition will become so complex that we have to realize that we are on the wrong track, at which point we give up and throw out the definition wholesale, beginning anew from a different angle. We make a qualitative leap, a transcendence.

The important lesson here is that only by making these attempts at definition, these quantitative changes, will we become able to see and take that qualitative leap. Only by trying to make the tacit explicit will we participate in a process of developing quality.

8

Culture

Quality Management
Control and Creativity
Culture and Structure
Strategies for Change
Engineering and Tinkering

Listen to the French philosopher and anthropologist Claude Lévi-Strauss telling us about an intervention into the Bororo culture, on the central Brazilian plateau: "The circular arrangement of the huts around the men's house is so important a factor in their social religious life that the Salesian missionaries in the Rio das Garças region were quick to realize that the surest way to convert the Bororo was to make them abandon their village in favour of one with the houses set out in parallel rows. Once they had been deprived of their bearings and were without the plan which acted as a confirmation of their native lore, the Indians soon lost any feeling for tradition; it was as if their social and religious systems . . . were too complex to exist without the pattern which was embodied in the plan of the village and of which their awareness was constantly being refreshed by their everyday activities."

Let us, for a while, imagine ourselves in the role of the Salesian missionaries but not confronted with the Bororo culture. Instead, we are hired by a big software house as quality consultants. Our task is to design a strategy for improving the quality of the organization's systems development projects. Now, can Lévi-Strauss be of any help to us in this effort as quality consultants?

Quality Management

At our introductory meeting in the software house, management informs us about the present situation and the background for asking us to intervene. Five years ago, the company invested a considerable effort in introducing a new quality organization based on a number of initiatives. Surveying the situation today, the initiatives have not, as seen by management, had the intended effect.

Some initiatives were concerned with the software house as a whole. A quality policy was formulated and a new quality assurance (QA) function was established with direct reference to top management. The QA function is responsible for developing general advice, guidelines, and standards on how to exercise the quality policy in specific projects. It is also the task of the QA personnel to serve as consultants for projects and to ensure that their practice complies with the quality policy.

Other initiatives have been concerned with specific projects. All projects have to develop their own quality plan in accordance with the overall quality policy and the general guidelines and standards. Each quality plan describes how quality is planned for in the project, how the quality of documents, prototypes, and systems is measured and evaluated, and subsequently how corrective actions are taken. The quality plan of each project has to be approved by the QA function, and during the project the QA personnel are engaged in evaluation procedures at important baselines.

This quality organization is drawn from the standard software engineering literature. Yet, management is quite dissatisfied with the present situation. There is a widespread impression that the actual operation of the projects has changed only a little, if anything at all. The marketing department has not experienced any noticeable change in the reactions from customers. There is still a severe backlog of requests from customers, and management has recently had to kill a project that went out of control.

After the introductory meeting with management, we spend some time talking to the QA personnel, and we have the opportunity to look closer at their activities. Apparently, the QA personnel have produced several guidelines and standards. The overall quality policy and the guidelines and standards are available in the company quality handbook, and during the last couple of years all systems developers have participated in a three-day seminar on quality control. The QA function has also succeeded in having quality plans designed in all new projects, and these plans are always reviewed and approved by the QA personnel. The QA personnel have spent a considerable effort producing the handbook and arranging the seminars. The seminars have been well received and the QA personnel are now engaged in updating the handbook and supplementing it with further guidelines and standards.

The QA personnel do not share management's pessimism. They believe the quality organization to be quite a success and quite advanced in comparison to those of other software houses. The QA personnel are often called upon to participate in open conferences and seminars to present the quality organization of the company.

The QA personnel admit that improvements are still needed. In their view, the major problem is the actual practice in the projects. All projects now produce good quality plans. But the systems developers put too little emphasis on following their own prescriptions. They tend to emphasize other aspects of their work. The QA personnel believe that the discipline in the projects needs to be improved. They would also like management to put more emphasis on the quality policy and to support the QA function in enforcing quality standards.

As we move on and engage ourselves in the views and beliefs of the systems developers, the picture becomes more differentiated. Some of the systems developers consider the introduction of the quality organization a big improvement. Others consider it a bureaucratic arrangement that is more of a burden than a help. Still others believe it makes no difference – except for the project manager who has to design the quality plan.

Quite a number of systems developers seem to have a negative attitude toward the quality organization. When asked to explain why, they give a number of reasons: The quality control system introduces a lot of extra administrative work; the quality policy and related guidelines and standards are merely formal expressions of what is known by any systems developer to be sound practice; the QA function is an expression of management's lack of confidence in the systems developers and their sense of responsibility. In addition, many systems developers believe that the QA personnel have developed a theoretical attitude toward systems development, and that they no longer have any feel for the practical problems involved.

There are, of course, variations between projects. But in general there seems to be quite a distance between, on the one hand, the actual practice in the projects and, on the other, the quality plans and the prescriptions formulated in the quality policy and in the related guidelines and standards. One difference is particularly conspicuous: the prescribed formal review procedure is seldom performed.

The quality policy stresses the importance of having persons independent of the producers of a computer system review all key documents during analysis and design. Instead, the projects tend to have the producers organize and conduct structured walkthroughs where other project members give their comments in response to the author's presentation. If reviews are in fact made, they tend to be less formal than intended. Project members rather than independent outsiders serve as reviewers. The authors are allowed to comment on the critique of the reviewers during the meeting. And very seldom is a strict acceptance procedure enforced.

Control and Creativity

The situation in the software house is in no way unusual. We are facing the classical contradiction between control and creativity. Focusing on the review issue, we should ask ourselves whether it is at all possible to control and manage creative activities to achieve better quality. To what extent were what we today consider the best of music, art, and fiction a result of individual creativity? And to what extent were these classical works a result of an externally controlled and explicitly managed process?

Systems development is a creative activity, and many systems developers see themselves as professional individuals, conscious and proud of their personal working style. The motivation and drive of many systems developers are founded in their practice as creative individuals. Systems development is, however, at the same time a highly cooperative and complex activity. Effective coordination is essential and the limited intellectual capacity of the individual systems developer has to be dealt with explicitly.

This contradiction is well expressed in Gerald Weinberg's notion of *egoless programming*. Reading and criticizing programs written by others is an effective way to improve the quality of programs, and at the same time it improves our competence as programmers to read programs developed by others. But systems developers tend to defend and protect their programs. This observation made Weinberg argue in favor of egoless programming: the idea that the individual systems developer should try not to view what she produces as being an expression of her ego. Instead, she should distance herself from her programs, and be open to criticism and intervention from others.

Most assembly line workers probably develop an egoless relation to the thousands of identical objects passing each day at close range in front of their eyes. But is it possible to develop an egoless relation to a creation of your own while at the same time being committed, involved, and creative?

Romantic philosophers like to stress the importance of self-realization, of opportunities for expression in work, art, and social relations. The identity we present to the world is the sum of our expressions, be they programs, touchdowns, hairdos, potatoes, or children. And the identity we present to others is the identity we will perceive ourselves to have. When we worry about who or what we are, when we are uncertain about ourselves, it is because our relation to our expressions is disturbed. Egoless programming invites such uncertainty of self, such alienation. A bit of alienation will probably increase our creativity, and there is something to the surrealist idea of "automatic," egoless poetry and painting, but most of us need to get our egos involved in order to care.

In many systems development methods we attempt to resolve this contradiction between control and creativity. Chief programmer teams and cleanroom development constitute two controversial examples addressing the same dilemma we see expressed in Weinberg's notion of egoless programming.

The basic idea of the *chief programmer team* is to divide labor between designers in a way that allows for individual creativity while at the same time enforcing explicit control. One person is assigned the role of the chief programmer and the majority of the team is programming under his supervision. Two other special- ized roles are the librarian, administrating documents and pro- gram libraries, and the assistant chief programmer. The chief programmer designs the overall structure of the computer system and programs the key modules himself. The rest of the modules are distributed to other programmers. The creativity of the individual programmer is restricted whereas one person, the chief programmer, is given maximum freedom to design the system.

Cleanroom development is based on a similar idea, enforcing a specific division of labor on the development process. Design and programming are separated from testing. Two independent groups are responsible for these tasks, and the design and programming group is not allowed to execute programs. Pro- grammers have to rely exclusively on program descriptions. They use specification techniques, editors, formal verification tech- niques, and reviews. But they are not allowed to run their own programs. The freedom of the programmers is restricted in this way to enforce a more strict discipline on the construction and description of programs.

Strict control and discipline are definitely needed to devel- op high quality software. But both of these methods, and many others, are controversial because they restrict the freedom and creativity of individual programmers. It is worth noting that much of the art produced in Europe before the nineteenth century, before romanticism, was produced using methods like these. Rembrandt was a chief programmer, the master of his workshop. But Rembrandt's society was an authoritarian one in which the notion of masters and disciples was natural. Even if we can see that such authoritarian technologies improve the productivity and quality in systems development, would we agree to use them ourselves?

Maybe we would try them once or twice. Maybe we would recommend that others use them. But would we really, if we had the choice, use them consistently and without modification over longer periods of time in our own work? Would we not try to modify them in ways that would maintain as much as possible of the intended discipline while at the same time making room for more individual freedom and creativity?

Culture and Structure

Our job as quality consultants is to increase the quality of the systems development processes and products of the software house. We are expected to have an informed opinion about the strengths and weaknesses of different software quality methods, and we are expected to know which methods to apply in which situations. Isn't it about time we start applying some of those methods? We cannot forever walk around talking with management, QA personnel, and systems developers, listening to what they have to say about quality and speculating upon general dilemmas involved in software development. If we want to produce change in the software house, why don't we get down to business and start practicing our religion? Certainly, the Salesian missionaries did not spend much time listening to the Bororo Indians. Maybe we have something to learn from those missionaries, after all.

We should not make the mistake of dismissing the Bororo example by reference to the simplicity of illiterate cultures like the Bororo. We now know that the cultures and societies we used to call primitive are extremely complex and difficult to understand.

In our own society much of the complexity is made possible by the printed word, by books. Our legal systems, political systems, organization of work, science, and technology would be impossible without books. Thus it is no wonder that we used to believe that illiterate societies had to be simple and primitive. More than anyone else, Lévi-Strauss has contributed to correcting our misunderstanding. Language, says Lévi-Strauss, is a system of signs. Written language is important as a means of communication because of its permanence. But all artifacts function as signs, as means of communication. Tools, methods, art, clothes, food, religious objects, buildings, family relations – all are signs communicating the social order from generation to generation.

When artifacts are thus seen as signs, as a means of communicating and supporting the social order, they gain importance beyond their material functionality and aesthetic beauty. Computer systems for communication, control, decision support, word processing, or whatever, do not only change the nature and division of labor. They also symbolize and communicate a social organization. The same holds for quality control systems, quality plans, systems development methods, project models, debuggers and other kinds of computer-based tools.

What we call cultures are complex phenomena composed of material structures (buildings, tools, machines), social organizations (families, relations of production), and systems of ideas (myths, ideologies, science). Based on his studies of foreign cultures like the Bororo, Lévi-Strauss is saying that material structures are symbolic manifestations and that social organizations and systems of ideas depend on the support of such material structures. In the Bororo village, the huts were placed in relation to one another and to the river, the forest, the men's house, and so on, so as to signal the status, wealth, occupation, and so on, of its owners.

A crucial element in the missionaries' strategy was to make the Bororo Indians abandon the circular arrangement of the huts around the men's house in favor of a village with the houses set out in parallel rows, as they are in an American suburban development. Once they had been deprived of their bearings and were without the plan which acted as a confirmation of their native lore, the Indians soon lost any feeling for tradition. By removing a fundamental structural foundation for the Indian culture, the missionaries created an opportunity for instilling a different set of cultural norms and values. This material intervention made it possible for other interventions to actually change the religion and life style of the Indians.

Lévi-Strauss is a materialist: To change a culture, you have to change its material structure. You can try changing its ideas or its social organization directly, but unless this change is supported by a corresponding change in the material structure, you can rest assured that all your efforts will come to nothing.

Applying these ideas of how to change a culture to our case, we first notice that there is a strong analogy between the situation in the software house and the situation in the nursing case discussed in chapter 6. In both cases a decision had been made for structural change, and a lot of effort had been put into implementing the decision. But both organizations remain stuck in a tradition, or company culture, which is very different from the culture we wanted the structural change to introduce.

In the nursing case we distinguished between three levels of organizational practice: the formal organization and procedures, the opinions of the nurses, and their actual behavior together with the deeply rooted assumptions governing this behavior. One of the lessons was that the initial change effort merely addressed the formal level without taking into account the opinions and actual practices of the nurses.

A similar pattern can be observed in the software house. A

lot of effort has been put into changing the formal organization and procedures to improve quality. A quality policy has been formulated, a new quality assurance (QA) function has been established with direct reference to top management, a quality handbook has been made available, and all projects have to develop their own quality plan in accordance with the organization's quality policy and guidelines. Nevertheless, there is quite a difference between this official and formal level of organizational practice and the opinions and actual practices of the systems developers within the software house.

In Lévi-Strauss' terms the previous initiatives have addressed the material structures without any real concern for the social organization and ideology in the software house. To improve the quality of the organization's software projects, we have to change the culture, and to do that, we have to change its material structure. But not as an end in itself. Instead we should carefully design structural interventions to create opportunities for continued social and ideological changes. If we just burn down the village and then walk away, the Indians will be sure to rebuild it.

Significant resources have been spent on improving the quality of software projects. Also, the initiatives have been based on sound principles of software quality. The previous efforts failed not because of limited resources or knowledge about software quality but rather because of the limited learning taking place in the software house. As consultants, we must succeed in addressing and changing the opinions and basic assumptions governing the behavior of the systems developers. Otherwise very little will be achieved. To identify and change the basic assumptions of the culture, we need to disturb the established routines and actively involve the systems developers in developing and adopting new professional standards.

We don't have a general cure for organizational ills irrespective of what is wrong with the organization. In this respect we are very different from the Salesian missionaries. What we have is our experience of working with organizations and a willingness to listen and let the patient talk. We are not interested in changing the symptoms but in getting at the mechanisms responsible for the symptoms. We are trying to change the systems developers' way of working, and we are engaging them in a process in which they will take responsibility for that change. We don't have a gospel to teach but are instead willing to listen and learn. The outcome is by no means guaranteed: Perhaps we will become Bororos rather than they Christians?

Strategies for Change

The Salesian missionaries want to turn the Bororo Indians into good Christians. This is indeed a dubious and ambitious goal that is open to criticism. What rights do missionaries have to enforce a new religion on other people? What is the actual effect on a culture when a new religion is enforced? Is it possible to import a new religion into an existing culture, or will the old culture just vanish? In the next chapter, we will discuss such questions as they are raised by the introduction of computer systems into organizations, but here we will concentrate on the practical lessons that could help us in our job as quality consultants.

We have no Bible, no fixed solution, in our pocket. We base our first attempts at intervention on our general ideas about software quality combined with what we learn about this specific software organization. We know that in order to change the organization we must understand it, but we also know that only by trying to change it will we come to really understand it. We have a general religion but we develop it further and adapt it to the situation at hand. Let us briefly summarize our general view on software quality, and afterward combine it with more specific strategic considerations.

Explicit specification of requirements is a necessary element in the struggle with quality. It is a practical and powerful approach to support communication and negotiation between users and systems developers. It is a useful approach to make the systems developers focus their attention during the actual development work. And it is sound practice to ensure that intermediate products are evaluated so that corrective actions can be taken as early as possible.

But we must also appreciate the subjective nature of quality. We should not naively insist on a practice in which all qualities are explicitly specified together with criteria and procedures for evaluating them. There is a substantial difference between quality in use and quality according to specifications. It is our ambition to develop computer systems that prove to be useful in practice, and this cannot be achieved even through the most careful systems construction approach. On the contrary, vigorous efforts to define explicit quality criteria will often be counterproductive. Quality can be obtained only if we are daring enough to trust our own judgments as well as the judgments of users, while still not underestimating the role of explicit criteria.

Systems development is a creative activity, and professional systems developers see themselves as individuals, conscious and proud of their personal working style. But systems development is also a highly cooperative and complex activity requiring discipline and independent evaluation procedures. Creativity is required to produce good software. But independence is required to ensure good evaluations. Effective quality control requires a certain division of labor and responsibilities. In practice, quality is not the only concern, and there is a constant struggle between quality and resource interests. Independence is needed to constantly defend a quality position and to avoid the self-deception involved in having systems developers evaluate their own products.

With such a religion on the one hand and our intuitions about the situation in the software house on the other, we are ready to outline a strategy for intervention. We choose to base our intervention on the assumption that the present difference between organizational formalities and organizational practices, and between individual opinions and practices is a rich source for organizational learning. There seems to be little need, in this software house, for new ideas about how to practice quality control. Most of the previous structural interventions were based on sound principles. The key challenge is to select and emphasize those initiatives that are useful and would have an effect in the present situation, to implement these into all levels of organizational practice, and to use this implementation to start a process of cultural change.

Our strategy will be, first, to have the software house make a decision to be formally certified according to a relevant standard – for example, ISO 9000. Such a decision will initiate a series of interventions that are different from previous experiences and that might make possible a change in fundamental arrangements, norms, and attitudes in the software house. Certification will, for instance, require that external institutions and actors become involved. ("Let us call in some missionaries!") The software house will have to describe and verify its quality procedures for the certifying institution, and the process might also require that external consultants are engaged.

An important reason for being certified is to gain competitive advantage in relation to specific markets, and this should motivate and commit the systems developers, the QA staff, and management more than yet another in-house change effort to improve quality and productivity. Thinking of a useful analogy to the Salesian missionaries' material rearrangement of living accommodations, certification might be the trick needed to create a breakthrough in our efforts as quality consultants.

We expect members of the organization on all levels to react and eventually engage themselves in a certification effort. Certification requires that quality procedures are described and followed. We have to find a way to produce a description of a set of procedures and a way to ensure that they express actual practices rather than formal arrangements and opinions only. The whole idea behind certification seems to be well suited to cope with the difference between the present procedures of the QA function and the practices of the projects. We must ask management to commit itself to this effort, and we must create a cooperative approach involving both QA personnel and systems developers.

The second element in our strategy will be to assess the maturity of the software house in order to select those key areas that can be most profitably addressed when we want to improve its software development operation.

Like people, organizations can be said to differ in level of maturity, and determining an organization's maturity level is not particularly difficult. For example, using a taxonomy such as the one recently suggested by Watts Humphrey, we would distinguish five levels of software-process maturity: initial, repeatable, defined, managed, and optimizing. Studying the software projects going on in the organization, this framework would direct our attention to the defining characteristics of the different levels.

Without going into detail about all five levels, a typical indication of an organization being, say, on the initial level of maturity is that people start coding whenever there is a crisis in their projects. On this level there might be formal procedures for various activities, but the organization lacks effective mechanisms to make sure that these procedures are followed. Projects are different and there are no systematic attempts to reuse and repeat successful patterns of behavior from one project to another.

The maturity model provides practical advice on how to improve the software development process – that is, on how to become more mature. To go from the initial level to the repeatable level, the organization has to design and implement basic management control of all development activities. In particular, it has to implement effective mechanisms for change control, quality evaluation, project management, and management oversight. Other key areas are suggested by Humphrey's model for further improvement of the software operation, in order to reach higher levels of maturity.

This maturity framework might not apply to our specific software house in all its details. But it provides us with practical guidelines for assessing present practices and for designing relevant interventions. In combination with the certification process, this analysis could provide valuable insights into which areas of improvement to start with and which to pursue later. Also our assessment of the present level of maturity can help us estimate the resources and time needed for the organization to be successfully certified.

Active involvement of managers, QA personnel, project managers and systems developers is crucial to our success. The quality issue is fundamental for all members of the organization. We need to take their different perceptions and interests into account to actually change the professional culture. The active involvement of the different groups of actors will motivate them and create more energy in the change process. Cultural change is threatening to an organization, and it should not be undertaken without understanding and respect for different perceptions and attitudes. To go smoothly, it requires the active participation of the involved groups of actors. There is, however, no guarantee of success. Latent conflicts may emerge and develop, and established power relations may dominate and make change difficult.

The certification process will make explicit some of the opinions and basic assumptions of the organizational culture, and this will produce new opportunities for useful interventions. In this way, we try to create an iterative process, in which actors reflect on the present situation, identify opportunities for interventions, and act to produce changes. This reflection-opportunity-action circle is the backbone of our strategy.

At this stage, we have only general and short-range plans combined with long-range goals. As soon as management has decided on a strategy, we can proceed and draw up more detailed, long-range plans. We will have to consider various forms for organizational learning. Workshops have to be arranged, experiments have to be conducted, and present procedures have to be evaluated, modified, and documented.

Engineering and Tinkering

Let us take this opportunity to leave the practical struggle with quality in the software house, and instead review the experience of the Salesian missionaries one last time. At first sight, Lévi-Strauss seems to argue for an extreme materialism, for a belief in the importance of material artifacts, technology if you like, in shaping social life. The Bororo Indians example could support a belief in technological determinism, the idea that technology determines the social order, having irresistible social consequences.

But Lévi-Strauss is no technological determinist. A new village structure, a new quality control system, a new systems development method, or a new computer system is not introduced into a material void. These artifacts take their place among the other artifacts in the organization, they are adapted to the way systems development is practiced, and their role in the organization will depend on where and how they are introduced. More profoundly, their role will depend on the way we think about them and the kind of cultural learning that will take place as they are taken into use.

There is a profound difference, says Lévi-Strauss, between thinking in modern societies and thinking in illiterate societies. Modern societies have engineers; illiterate societies have *bricoleurs,* or tinkerers. As engineers, we organize our thinking in projects, choosing means and tools once the aim of the project has been decided. As tinkerers, we use what we have, letting our means and tools determine what we do. As engineers, we set our goals first, often having to invent tools to be able to reach them. Our projects determine what tools we have. As bricoleurs, we stay safely within the world of tools we have acquired. Our projects are made possible by the tools we have. The tools we have are the ones we picked up along the way because they might come in handy.

Engineering means developing expert tools, tools that are optimally designed for their task. Tinkering, on the other hand, means using whatever happens to be available that will at least do a satisfactory job. One and the same tool will have to play many different roles in its life. This has the consequence that we, as engineers, are trained to optimize, while as bricoleurs we are trained to satisfice. As engineers, we are trained to see function first and form only later, to construct forms, top down, for a given function. As tinkerers, we are trained to see form first and function later, to discover, from the bottom up, new functions for a given form. An engineer will look for the function of an artifact, not really seeing the form when the function has been determined. A tinkerer will look closely at the form, remembering its details, in case the artifact one day may have to play a different role.

Now you see what the notion of bricolage does to our calling Lévi-Strauss a technological determinist. If we deal with computer systems and systems development methods as engineers, we see only the function they were designed for, and the artifacts will indeed determine our social organization, be it our own software organizations or the user organizations. But if we deal with computer systems and systems development methods as tinkerers do, we will see in them whatever we need to see that is at least possible to see.

If we are tinkerers as well as engineers, there is virtually no limit to how different the social orders can be that are supported by one and the same computer system or systems development method. But this does not alter the fact that when an artifact supports a certain social order, that order is threatened when the artifact is changed.

By changing the material artifacts, computer systems, and methods of an organization, we can threaten its social order and its ideology, both its organizational structure and its culture. But if Lévi-Strauss is right, we have very little control over the ensuing change unless we, like the Salesian missionaries, complement our technological intervention with an ideological one. Left on its own, the organization will turn its tinkering capacity loose on the technology with a result beyond our wildest anticipation.

Observing only how differently our colleagues work with the very same computer systems or programming methods, we should not be surprised at how differently computer systems may turn out in the hands of different organizations. When we begin to view the world in terms of tinkering rather than engineering, we begin to understand why it is so difficult to plan for quality.

Why did the software house decide to call upon us as quality consultants? Obviously, they had already spent a lot of time and energy trying to improve software quality. What did they do wrong? Maybe they approached the quality issue as engineers, relying on superficial organizational interventions and neglecting the deep ideological challenge. Maybe they practiced too much thinking and too little tinkering.

Improving the quality of systems development projects and of computer systems is a struggle. The introduction of new forms of organization and of new methods and advanced tools are examples of structural interventions that might facilitate cultural change. But such interventions are in themselves insufficient. Unless we really engage the competence and basic assumptions of the individual users and systems developers, our cultural change efforts will come to nothing.

9

Power

Computers at Work
Expert or Political Agent
Interests
Artifacts, Culture, and Power
Artifacts Have Power

Telephone services change with computers. The telephone network used to contain a great number of manually operated switchboards. Today, these switchboards are automated. Information about telephone numbers and addresses has traditionally been compiled in telephone books. Now, this information is available on electronic media. When you dial the telephone company, an operator uses a computer-based phone book to assist you. When telephones are integrated with terminals, this kind of service is directly available to each individual customer.

In one case, a nationwide telephone company invested in a new computer-based information system to provide its customers with more effective services related to phone numbers, names, and addresses. Instead of developing its own computer system, the company decided to buy a modified version of a system that was already in use in other countries. A contract was made with the foreign software house that had developed the system. The contract contained a detailed specification of the new computer-based information system. Some parts of the old version of the system had to be modified, and new facilities were to be added.

The telephone company wanted to have the monitoring of operators – a special terminal making it possible for a supervisor to monitor each operator by inspecting a copy of the operator's screen – included as a new feature. One of the programmers in the foreign software house did not like this idea and decided to add one extra feature: When the supervisor inspected the screen of a specific operator, a sign would appear on the operator's screen to indicate that it was now being monitored. This facility was not part of the requirements specification, and the telephone company later had it removed.

Was the programmer right in trying to sneak in extra features to create, in his view, a better computer system? Who is responsible for the impacts of a given computer system on the quality of work and the quality of life? Who should have the right to influence design decisions?

Computers at Work

Computers and products containing computers now appear in a huge variety on the market. Just think of personal computers, computer games, household machines, cars, and communication technologies. We buy and use these products, and they influence the way we live. At the same time, all these computer-based products require new or modified production processes and they deeply influence working life.

At the workplace, computers are used to automate work. Tasks that used to be performed by humans are now performed automatically by computers. These tasks are either information-processing tasks, like accounting, or control tasks related to other processes, like the regulation of machines or chemical processes. Using computers in this way changes the relation between people and technology at the workplace, and it raises fundamental issues related to employment, division of labor, and work organization.

Computers are also used to plan and control work. Any organized human activity needs planning and control, and this in turn requires information. Computers are widely used as tools to manage work processes and to economize human resources, technology, and materials. Computer-based management systems have an impact on the way work is organized and on the way individual tasks are performed.

The use of computers, or any technology, affects employment. For a given task, increased automation reduces the amount of labor needed to perform that task. From this local point of view, we would expect computer usage to lead to unemployment. The relation between computer usage and employment is, however, much more complex. New jobs appear as a result of increased production – that is, as a result of economic growth. New jobs are also created by the introduction of computers, and, as a side effect, we experience increased employment within other branches and departments.

When computers are introduced, or when existing computer systems are modified, the established work organization is challenged and modified. Any established division of labor is intrinsically related to the use of technology, and very rarely is it possible to change the one without changing the other. As traditional machines on the shop floor are replaced by numerically controlled machines and later by computer numeric machines, the division of labor is challenged. The new possibilities to program each machine affect the division of labor between the production planners and the shop floor. On the shop floor new boundaries are created between skilled and unskilled work. Traditional machines require constant monitoring and regulation by skilled workers, whereas computer numeric machines require less regulation and more routine monitoring.

The use of computers affects the very content of work. One of the key aspects of any job is the extent to which a worker participates actively in shaping and planning it. As computers are used extensively to plan, control, and manage work, they affect what is considered by most people to be crucial elements of a good job. Some functions disappear as computers are introduced, either because they are automated, or because work is divided differently. Other functions change, and new functions emerge.

Last but not least, the use of computers influences the ergonomic and physical conditions at the workplace in positive as well as negative ways. Using computers is very different from using paper, pencil, and typewriters. People spend hours in front of the screen, working in the same position, and they communicate via electronic mail, which means less walking around. On the other hand, using robots to manipulate objects in front of heated ovens or to automatically control the painting of objects removes humans from unhealthy and dangerous working situations.

There are always choices related to the ways computers are brought into use. But the ways we use them have, in any case, a profound influence on individuals, organizations, and society. It is not surprising that fundamental political issues and difficult moral dilemmas are raised by the use of computers. This again raises a practical question: What can and should we as computer professionals do? In what ways are we, as professionals, responsible for how computers are used?

There seem to be two fundamentally different answers to these questions. We can take the position that, as computer professionals, we are responsible for developing good systems that satisfy the needs of our users. Or we can take the position that the use of computers is the responsibility of the users themselves, and that, as computer professionals, we should support them in developing their competence so that they can make their own choices.

But this is an academic way of expressing the issue, relying as it does on an unanalyzed use of the notion of user. Before we can decide whether we or the users should be responsible, or rather, how we should divide responsibility between us, we must identify the user. Who is the user in the telephone company example above? Is it the company management who orders the system and who wants the monitoring device? Or is it the telephone operators who will have their work monitored if the device is installed? Obviously, these two groups have different interests. To satisfy them both at the same time is impossible. A good system for one group is not so good for the other. As long as we ask only who is responsible, the user or the systems developer, we presuppose that there is fundamental consensus among the users of the system, not to say among all the people affected by the use of the system. When we realize that this is not the case, we also see that another question is more pressing: In whose interests should we develop computer systems?

These two positions – naive belief in the shared interests of all users and realization that there are conflicting, contradictory interests among the people affected by a computer system – correspond to the attitudes of what we will call the expert and the political agent. Experts, typically, develop their systems for an abstract and idealized user and can thus remain blind to a world of conflicting interests with limited doses of self-deception. Political agents make a deliberate decision about whose interests they want to support in their work. This decision is normally implicitly derived from what it takes to remain in business and make a career or profit.

Expert or Political Agent

Traditional *computer experts* are offering technical expertise, and they expect users to make explicit and specify what they want them to do. They know about computers and how to build computer systems. The users are expected to know their own business, and they should be able to evaluate the use quality of a specific computer system. The attitude of traditional computer experts is humble and comfortable. Their expertise is limited, and it is the responsibility of others to use it. The weak point of this attitude is that its effectiveness is questionable. Is it really possible to separate concerns in this way?

Some experts have a slightly different understanding of their responsibility for the ways in which computers are used. They see themselves as sociotechnical experts rather than as computer experts. *Sociotechnical experts* believe that only by taking social issues into account will they be able to develop good systems and get satisfied users. To them, social issues are as important as technical ones – sometimes more important. They believe that the computer is only a potential solution and that the real problem is of a social nature. Still, with their sociotechnical methods of social and organizational analysis, they very much act like experts.

As experts, they have to understand and appreciate the organizational setting. They have to know what kind of information processing is required, and they have to analyze and appreciate the attitudes and expectations of the future users. In doing this they apply a variety of techniques, such as analysis of variance and job satisfaction analysis. In addition, they perform a more traditional, technically oriented analysis, emphasizing data and computer related issues.

A sociotechnical approach requires the active participation of users. Experts need the cooperation of users in order to really appreciate their users' needs and expectations, in order to understand their organizational context. The attitude of these experts is concerned and cooperative. They master technical as well as socially oriented techniques, and they know how to design and manage cooperation with users. But like traditional experts, they consider themselves to be experts confronted with problems generated by the users.

The situation is assumed to be fundamentally harmonious. The sociotechnical expert does not worry about who the users are and whether they have the opportunities and resources needed to cooperate and say what they want. So, of course, the most commonly raised criticism against the sociotechnical expert is that this approach often has the consequence that users are taken as hostages. Maybe the weak point in the sociotechnical program is that it promises too much.

Political agents see the situation differently. They argue that as professionals we have to commit ourselves and answer a fundamental question: Who really is the user? This question will often have a complex answer, and depending on the answer the world will look very different. But having committed ourselves we will be able to assist a specific group of users in developing their resources, insights, and technical support, thus increasing their power.

The political agent cannot say "good" without asking, good for whom? The evaluation of computers in use is always relative to one or another interest. People with different interests will not be able to agree on the quality of a computer system. Therefore it is only meaningful to try to develop good systems together with users that share basic interests. As computer technology has been introduced into newspaper production, substituting electronic technologies for a lead-based production technology, work has changed dramatically. The political agent would argue that in such a process, management, journalists, and graphic workers have contradictory and even antagonistic interests. No matter what kind of expert you are, you will not be able to develop computer systems, even in cooperation with users, that all involved interest groups will accept as being of high quality.

Management wants to utilize computer technology to develop a more effective production process and to become less dependent on graphic workers. Journalists want to use the opportunities provided by the new technology to take over some of the functions that were traditionally carried out by graphic workers – for example, by performing layout and make-up in addition to editorial functions. Graphic workers, however, want to preserve and improve their own skills and position at the newspaper.

Political agents would in this case have to make a decision. If they choose to cooperate with the graphic workers, they would have to design computer systems based on principles protecting their interests. Such a design would be based on principles like graphic quality of products and processes in newspaper production; democracy at the workplace with respect to the ways work is organized and managed around the computer systems; and further education to support local development and improved graphic skills.

The resulting computer systems would not allow the journalists to take over functions that have traditionally been carried out by graphic workers. The systems would maintain and further develop the idea that graphic competence is different from journalistic competence and that this kind of competence is needed in its own right and is to be distinguished from journalistic competence.

The computer expert develops technology. The sociotechnical expert works with people. The political agent intervenes and takes sides in a power struggle. Historically, systems development was first done by computer experts, and most of it still is. Sociotechnology is very much a British invention that through ideas such as user-participation and social design has had some influence on how systems development is now being done.

As we define "political agent" here, the notion applies both to the typical consultant wanting to get ahead and make a profit, and to the radical left-wing consultant coming out of the student movement of the late 1960s. In Scandinavia the latter position came to dominate systems development research at the universities for a couple of decades. Working with trade unions as opposed to management, taking sides in a struggle for power over the new technology, these researchers were very much political agents.

To such radical political agents, traditional computer experts seem naive but sociotechnical experts are hypocrites. Traditional experts have a much too simple conception of the situation. They are not concerned with the context of the computer system, and they ignore the organizational and political conditions for developing a new system. Sociotechnical experts have, in the view of radical political agents, a deeper understanding of the situation. But by promising quality for everyone they necessarily have to manipulate someone.

A criticism that has been raised against the attitude of the radical political agent is that it is really a conservative, rather than a radical position, in the sense of taking the traditional division of labor for granted. In deciding to defend the graphic workers' interests, for example, the political agent becomes blind to the possibility of fundamental changes in the division of labor as offered by new technology. Neither the technical nor the sociotechnical expert is similarly hampered.

Systems development projects are typically initiated and sponsored by management. Only research and development projects can be initiated and sponsored by other interest groups. Radical political agents try to do the impossible: They try to utilize the development and use of computer systems as an opportunity to revolutionize organizations and society.

We are all computer professionals. But some of us develop expert attitudes, others become political agents. Some of us continue to see ourselves as working with technology, while others discover that our subject is really technology and people. Some of us become political agents for management, others for graphic workers. What motivates these choices? And are they choices?

Most of us become what we are without really knowing, or caring, how it happened. After the fact, we construe values and explanations that convince ourselves that we are doing the right thing and for good reasons. We are more interested in defending what we have become than in questioning what we are. And we dislike such questioning when it is delivered as criticism, as it so often is. Who wants to have his identity questioned by someone else? Better then to do it yourself by philosophizing upon the different positions possible within systems development.

Such philosophical questioning can be pursued from within, relying on the professional experience we have gained as systems developers. But it can be carried further by drawing on the theorizing of philosophers who make it their profession to question and compare different positions, even if those philosophers seldom explicitly address the profession of systems development.

Interests

The theory of different knowledge interests, developed by the German philosopher Jürgen Habermas in the 1960s, led to a rather heated discussion about the aims and goals of science. Not much to discuss, you might say, since no one in his right mind will object to knowledge, the aim and goal of science. But we don't pay for research unless it brings knowledge that we can use. And therefore the use we value will determine what sorts of knowledge science will be seeking. Even if the financing of, say, academic research won't control its details, it certainly will determine the general trends.

Knowledge is useful but it can be used in many different ways. Depending on the use we have in mind, our quest for knowledge will be motivated by different interests. Habermas discusses three such interests: technology, understanding, and emancipation. It is not difficult to apply Habermas's discussion to the goals and aims of systems development. The question "What should we use computer systems for?" is similar to the question "What is it we want scientific knowledge for?" Habermas' knowledge interests can serve just as well as systems development interests.

The natural sciences are dominated by an interest in control, according to Habermas. We use *induction* to arrive at causal laws that we can use to explain and predict events in nature. But why are we interested in explanation and prediction? Because we want to learn from our mistakes and act with foresight. "Why did you flunk your exam?" and "What do you think will happen if I just boot?" are questions aimed at making life easier. We use the answers to guide our actions, to control the world.

Think about the number of causal laws that every child must learn in a few years just to survive. It is awesome. How to open doors, turn on lights, get help, handle objects. . . . So, the next time you see a child repeatedly upsetting her glass of milk and intently watching the result, don't interfere. She is a scientist at work using induction to increase her knowledge about the regularities of nature, to enlarge her store of causal laws.

Only to the extent that the world is regular can we control it. Things must repeat themselves, there must be law and order, for us to be able to explain, predict, and be in control. Thanks to the regularity of the processes, we can turn our knowledge into a machine that does the controlling for us. The essence of methods, tools, and machines, of technology, is control. It is because the machine is such an outstanding instrument for control that it is such a good example of technology. We use technology to control the world. When we want to use knowledge to develop our technology, our interest is *technical.*

That the natural sciences are dominated by an interest in control is all right, says Habermas, but when this interest is exported into the humanities he objects. History is not interested in control; history is interested in understanding. The events that interest a historian don't repeat themselves. The causal laws that you can discover in history are terribly superficial and irrelevant to the complex events and processes that interest a historian.

With an interest in control, we ask why an event appeared, taking for granted that we know what event it was. The real work for a historian lies in identifying events, to say what really happened. Historians are happy if they can give a coherent description of their subject matter, making some sense of their data. Before they can even think of finding regularities between events, they must decide what the events were. A battle, a political crisis, a wave of emigration, an increase in crime rate, industrialization – what are they, when do they begin, when do they end, what do they include? To answer such questions, historians must *interpret* their data. Interpretation rather than induction is the method they must use.

History is made by people. Historians study the actions of people. Their ambition is to understand those actions, to make sense of them by identifying intentions and the cultural context in which the action takes place. They are happy to understand what happened, to find out what X did and why and how people reacted to the action. The humanities are interesting because there is something to be learned from others, from other cultures, from people in the past and from people around us. But what we learn from history will broaden our vision rather than help us get along with material objects. Our interest is *understanding*.

The social sciences can study social processes with a technical interest, with the goal of learning how to control them. Just as there is physical engineering, there is social engineering. Social engineering – that is, attempts to control society and people by the manipulation of causes – is criticized by Habermas for not perceiving the fundamental difference between society and nature. Since society is constituted by its members, there can be no social laws with the stability of natural laws. Social engineering will seem possible only to the extent that the members of a society are oppressed.

Anthropology, the study of foreign cultures, is often said to be motivated by an interest in understanding those cultures. But Habermas is not happy with understanding as the ultimate interest of social research. To study a society in order to understand it presupposes that a common understanding is possible. But as long as power is unequally distributed, our understanding of society will differ depending on where we are in the power structure, and all attempts to develop a common understanding will have to suppress our different interests.

Because of the unequal distribution of power, the social sciences cannot be guided by either technology or understanding. Until our societies become democratic, until values such as liberty, equality, and justice are realized, the aim of social research can only be to promote these values, and its interest has to be emancipation. Rather than collecting regularities for the purpose of control, we must analyze the regularities to determine whether they are oppressive or not, whether they are legitimate or illegitimate from a democratic perspective. Rather than developing a common understanding of society, we must analyze the common understanding of society, the ruling ideology, and criticize its oppressive components.

Our societies use their ideologies to cover up an ugly reality. Efficiency, rationality, freedom, competition, progress – it all sounds so impressive, but what it really means is a rat race where the weak, sickly, and handicapped are eliminated, where you work yourself to death in order to buy the latest gadgets, constantly worried about not performing well enough and losing your job.

Such *criticism* of ideology is, according to Habermas, the main task of a social science. By criticizing the ideology of a society, you prepare for emancipation. Oppression can be fought only by people who realize that they are oppressed. If you think you are born equal, or that being born equal is not for you, then you will not fight for your rights. Uncle Tom will not join the civil rights movement.

Technology and understanding are two fairly straightforward interests. Emancipation differs from technology and understanding in going beyond our common understanding of the goals and aims of science. Emancipation seems to take us out of the sphere of science into political action. But Habermas would argue that exactly the opposite is the case.

As soon as we appreciate the importance of the fact that all societies are characterized by the unequal distribution of power, we realize that a study of society makes little scientific sense if it is not motivated by the interest of emancipation. As long as it is motivated by a technical interest or by a desire to understand, the most important facts about society will remain hidden. Thus traditional science is really a form of political activity in defense of the established power structures. Only by choosing an *emancipatory* interest will our scientific research really aim at knowledge.

The whole idea of different interests expresses a belief that science, or in our case systems development, is motivated by interests or values. In our society, technology and understanding are so uncontroversial as values that we fail to see this clearly. But when it comes to using science or computers to change the relations of power in our society, when emancipation is put forth as a knowledge or development interest, then the question of values becomes more controversial. Who is to be emancipated and from whom? Who is to lose power and who is to gain? And how can it be the business of scientists or computer professionals to take part in a political struggle for power?

The more unequal the distribution of knowledge in an organization or a society, the less democratic the political process will be. But since knowledge is a resource both in production and marketing and in the exercise of power, there is a strong incentive in a competitive society not to make knowledge a generally shared resource. There is a conflict between a democratic ideal of enlightenment and a belief in the virtues of competition. But this conflict does not hinder us from believing in the importance of a free press and in keeping academic research in the public domain. In a democratic political system it is generally appreciated that knowledge is a resource needed for participation in the political process. Emancipation is, in this sense, uncontroversial as a knowledge interest in a democracy.

It is, however, one thing to believe in the importance for a democracy to make knowledge a widely shared resource, and quite another to let our research or practice as computer professionals be motivated by such a belief, seeking the kind of knowledge that will best serve such a democratic purpose. It is still more controversial, of course, to let our practice be motivated by an interest in emancipating one or another group or category of people that we consider to be severely oppressed. But is there really a difference between turning a belief in the value of technology into specific technical solutions to problems in a particular workplace, and turning a general democratic attitude into emancipating a particular group of people?

Artifacts, Culture, and Power

Habermas ties his knowledge interests to what he calls means or media of social organization. A society is based on three different means of organization: labor, language, and power. To change an organization or to change society, you have to act on these means of organization. Our views of quality in systems development and

qual·i·ty (k wŏl′ə tĭ), *n.*, *pl.* **-ties. 1.** a characteristic, property, or attribute: *useful qualities.* **2.** character or nature, as belonging to or distinguishing a thing: *the quality of a sound.* **3.** character with respect to excellence, fineness, etc., or grade of excellence: *food of poor quality.* **4.** high grade; superior excellence: *goods of quality.* **5.** native excellence or superiority. **6.** an accomplishment or attainment. **7.** *Archaic.* social status or position. **8.** good or high social position: *a man of quality.* **9.** the superiority or distinction associated with high social position. **10.** *Archaic or Dial.* persons of high social position. **11.** *Acoustics.* the texture of a tone, dependent on its overtone content, which distinguishes it from others of the same pitch and loudness. **12.** *Phonet.* the timbre or tonal color which distinguishes one speech sound from another and remains essentially constant for each sound, even in different voices. **13.** *Logic.* the character of a proposition as affirmative or negative. [ME *qualite*, t. L: m.s. *quālitas*]
—**Syn. 1.** trait, character, feature. QUALITY, ATTRIBUTE, PROPERTY agree in meaning a particular characteristic (of a person or thing). A QUALITY is a characteristic, innate or acquired, which, in some particular, determines the nature and behavior of a person or thing: *kindness as a quality, the quality of cloth.* An ATTRIBUTE was originally a quality attributed, usually to a person or something personified; more recently it has meant a fundamental or innate characteristic: *an attribute of God, attributes of a logical mind.* PROPERTY applies only to things; it means a characteristic belonging specifically in the constitution of, or found (invariably) in, the behavior of a thing: *a property of hydrogen, of limestone.* **3.** nature, kind, grade.

of how computer technology changes organizations and society will differ, depending on what medium we consider to be the more important.

If we think of organizations and society as based on *labor*, the production of goods and services, then we will think of changing society by changing the means of production and we will think of changing organizations as a technological challenge. Technology will be identified as the major social force. By developing technology, we change our ways of working, and we make our production of goods and services more efficient, more rational. Systems development will mean construction of computer artifacts, and quality will be closely related to the functional properties of those artifacts.

If, instead, we regard *language* as the most important means of social organization, then organizations and society will change with the changes in communication and social interaction. The relations of production rather than the forces of production will determine social life. Technology will be important to the extent that it influences patterns of communication within society. Systems development will concentrate on the culture. Artifacts will evolve in a dialogue with users, and quality will be related to the artifact in its context of use rather than to the isolated artifact.

Theories stressing the role of labor and artifacts tend to be future oriented, and they welcome technology-driven social change. Theories stressing the role of language and culture tend to be more conservative, seeing in technology a threat to a tradition-bound understanding. When society is seen as organized by relations of *power*, technology is viewed as weapons to be brandished by oppressors or the oppressed in the power struggle. The quality of a new technology is determined not so much by its functional, aesthetic, or symbolic properties, as by who has the power to control its use and by its influence on the distribution of that power.

If we use Habermas' three interests, and his media of social organization, to reflect upon the practice of systems development, we come up with three different professional positions: engineer, facilitator, and emancipator. These positions may at first seem to be just another version of the same old threesomes: construction, evolution, intervention; hard systems, soft systems, dialectic systems; or traditional computer experts, sociotechnical experts, and radical political agents. But the fit between these threesomes is by no means perfect, and Habermas' taxonomy of interests will enrich our understanding of what it can mean to say of a computer system that it has quality.

Traditional computer experts have a technical interest and want to improve the world by developing better computer artifacts. They are *engineers*, with knowledge that gives them superior control over the processes of computing. Sociotechnical experts may at first seem to be motivated by an interest in understanding and culture, but a closer look will reveal them to be dominated by technology. Sociotechnical experts are engineers, only more of social engineers, trying to control not only the processes of computing but the processes of human-computer interaction as well. The aim of the sociotechnical experts is not to increase their own, or the users', understanding of computer systems and their use. Their aim is to increase the efficiency of computer technology use. Understanding may very well be an important means to achieve this aim, but it is not the aim itself.

Systems development is, of course, dominated by a technological interest. It is difficult to see how understanding could be the final end of research or practice in our discipline. But we could assign to ourselves the role of *facilitator* rather than engineer, thus seeing our task as one of increasing the competence of our customers, handing over to them the responsibility of acting on the basis of that competence. Our responsibility would lie not in a better technology but in a deeper understanding of technology and its use. We would not be working as experts for a more or less abstract user, but as facilitators, teachers, and problem solvers, typically, with particular users.

Radical political agents are *emancipators*. Their interest is emancipation. They want to use the development of computer systems as an opportunity to advance society and social organizations in the direction of equality, justice, and democracy. Emancipators use a variety of approaches. Like the political agents working for management, as most of us do, they rely on construction, evolution, or intervention, depending on the role they think computer technology plays in the struggle for power.

Engineers want to increase the efficiency of computing and computer use. Facilitators strive to increase the understanding of how technology could be made to serve people rather than the other way around. Emancipators worry about injustice, blaming almost everything on an unequal distribution of power, often supported by the use of advanced technology. None of these positions need to be less moral than the others, less socially concerned. The difference between them lies rather in what is considered to be the most important factor to attend to if we want to improve the world: wealth, understanding, or equality.

In recent years Habermas has tried to develop his interest in language and power into a general theory of communication. Seeing rational discourse, argument, and dialogue as a prerequisite for a functioning democracy, Habermas has discussed conditions and obstacles for what he calls ideal speech situations. Three such conditions can be defined in terms of the three different worlds we as human beings are related to, according to Habermas. Communication aimed at understanding will have to be true of the objective world, right in the social world and sincere to the internal world. Again, there is a clear parallel between this taxonomy and the three interests.

If we want to use Habermas as an inspiration for our work as computer professionals, we can define quality, generally, in terms of ideal speech situations. A good project and a good computer system is such that it makes possible communication, understanding, and rational discourse. When we design a project or a computer system, we will take pains to make it a help rather than a hindrance in achieving ideal speech situations. And when doing so, we will aim not only at truthfulness but also at rightness and sincerity. We may have a tendency to view information systems generally as systems for communication. Expert systems, for instance, will be seen as communication systems relating human experts to clients, and judged accordingly. And our evaluation of the quality of the system will be divided into three dimensions: the truthfulness of the system, its rightness, and its sincerity.

We cannot judge the quality of an artifact without considering the interests involved in its use. So, in order to define our own quality standards, we have to reflect upon our development interests, what we develop computer systems for. We don't have to use Habermas to do that. We can use concrete examples.

Which factor is more important: that the systems you design challenge the user to understand them or that they are easy to use? The home computer of the 1970s was difficult to use to solve practical problems. It was in many ways hopeless as a piece of household technology, but it was a simple machine that invited users to develop their own programs in Basic. The home computer was advocated by some for this very reason: to help increase peoples' understanding of computers and programming. It was eventually to be superseded by a more user-friendly personal computer, equipped with all sorts of program packages ready for practical use. The personal computer is a much more complex machine, difficult to understand but demanding no understanding of programming. Now, is this progress or not? Which computer, the home computer or the personal computer, has the higher quality?

Or, let us take a different example. What is more important: to develop efficient technical solutions or to have satisfied users? When computer technology was first introduced for administrative use in the 1960s, there was much discussion of the dequalification of work brought about by this technology. It was claimed that the introduction of computer technology in this respect resembled earlier use of machine technology to automate production. The more interesting aspects of the work process were taken over by technology, leaving human beings to perform routine, boring, repetitive tasks serving the machine. If we are motivated by understanding or emancipation, these aspects of the use of information technology will be at the center of our attention as systems developers. If we are motivated by a technical interest, we will not worry about these aspects. Unless, of course, they interfere with the overall efficiency of the computer-based system. We will be frustrated by the resistance of the users to our clever technology.

Artifacts Have Power

Only the traditional computer expert really believes in technology as a positive social force. Both the sociotechnical expert and the political agent stress the importance of what goes on in the context of the technology. Traditional experts tend to argue that if only there is technical progress, the rest will take care of itself. They tend to be technological determinists. Sociotechnical experts and political agents both argue that the role of a certain technology will be determined by the social context into which it is introduced. They are social constructionists.

Technological determinism is the theory that a developing technology will have social consequences for us to foresee and then live with. As the earth is shaped by wind and water, society is shaped by technology. The experts designing our technology are responsible for the shape of our society. When Robert Moses built parkways on Long Island that were meant to take the automobile-owning, white upper middle class out to Jones Beach, he made sure that the overpasses were low enough to prevent buses from using the parkways. So, artifacts have power. Artifacts make up the world we live in, and they set the limits for our freedom, not only of movement, but of how much time we have, whose company we share – in short, what we do with our lives.

A very different type of theory views technology as a social phenomenon shaped by the society producing it. Technology is said to be *socially constructed.* This can mean simply that the social conditions of the design and development of new technology are being stressed. Or it may mean, more generally, that the quality of a technology depends on how people conceive it, what they know about it, their attitudes toward it, how they decide to use it. Technology is what its users perceive it to be. Such a social constructionism goes well with a democratic attitude toward the design of technology: We move from design for or with the users to design by the users themselves. Rather than being complications in a causal chain of engineering, the users turn out to be the real designers.

An appreciation of the idea that technology is socially constructed by its users changes our conception of systems development. Using computer technology to change an organization should not be viewed as a process of engineering. For reasons of democracy, such a perspective would be degrading, and for reasons of making a profit, it would be silly to so underestimate the complexity of social response to technical change. The heart of systems development can no longer be a product, a computer system, produced by professional systems developers, since the properties of that system will be determined in its use. Systems development becomes part of a wider and open-ended process in which the systems developers play a marginal role.

If we are technological determinists, we have to accept the responsibility as technology experts for changing the lives of people. If we are social constructionists, we share this responsibility with everyone who is involved in the design and use of technology. In the first case, we will have to worry about the quality of our systems, taking into consideration its far-reaching social consequences. Designing our systems, we shall have to include their future use and effects on conditions of work and life in our planning. In the second case, as social constructionists, we will be interested in engaging in the design process the users, and not only the immediate users but all who will be affected by our systems. But we will share this interest in making the design of technology a democratic process with our fellow citizens. In that process we are no more responsible than any other citizen, and as computer technology experts we can rest content with contributing to that process our technical expertise.

There is something of a paradox in this conclusion. Technical experts tend to be technological determinists. Not giving too much thought to the conditions of use, they develop their artifacts as if their use were determined by their functional properties. Unwittingly, they accept a formidable responsibility for the social consequences of their technology while at the same time refusing to consider these consequences. Social constructionists see more clearly the complexity of the interplay between society and technology, but in doing so they distribute the responsibility more widely, making it possible for themselves to stick to their role as experts.

There is some truth in technological determinism, and there is some truth in the idea of the social construction of computer systems. But looking back, it clearly seems as if the changes taking place over the last hundred years or so have been the result of technology shaping us rather than the other way around. We can formulate this as a theory of technology and social change and call it, with Langdon Winner, technological somnambulism: "The interesting puzzle in our times is that we so willingly sleepwalk through the process of reconstituting the conditions of human existence." People in general have had very little to say in the development of computer technology, and they have mostly remained ignorant of the options available. The computer professionals have not given much thought to the social consequences of their technology.

If we are dissatisfied with a society characterized by technological somnambulism – something computer professionals ought to be, believing as they do in planning, requirement specifications, and rational design – what can we do about it? As computer technology invades every corner of society, it changes in innumerable ways every aspect of modern life, including work, health care, education, and warfare. As consumer goods, computer artifacts will be used in automobiles and in the home, and by children in and out of school, for play and pleasure. As citizens, we share responsibility for these changes with every other citizen. Computer professionals are not more responsible than anyone else for these more general changes going far beyond their own minuscule contributions. But they have an expertise that makes them morally obligated to speak up against the development of low quality computer systems and the irresponsible use of computer technology.

The programmer mentioned at the beginning of this chapter did speak up. He tried to sneak certain features into the computer system to compensate for what he believed to be bad management practices in a telephone company. The project was established in such a way that there was no contact at all between systems developers and users. The programmer was sitting in a different country, basing his work on a contract between his software house and a foreign telephone company. His closest contacts with clients and users were the consultants in the sales department of the software house. He was caught in a situation in which cooperation with the users was impossible. In spite of this, he tried to modify the system to make it more human, as he saw it.

There seem to be at least two morals to this story. The first is that, like any professionals, computer professionals should be conscious and critical of the basic conditions for doing their job. The organizational setting, the contractual arrangements, the economic and political conditions influence and in many respects determine our practical possibilities for being concerned with quality, with the ways in which computers are used and the conditions we create for other people in their jobs.

The second moral is that independently of our basic attitude it is always important to ask the fundamental question: Who is the client and who are the users? How can we achieve quality in this project and what does that mean? If we are not seriously concerned with these questions, we have no chance of understanding the manipulations and power struggles in which we take part. We cannot develop a professional attitude toward our work if we shut out the reality in which we are involved, concentrating only on the immediate task at hand. On the contrary, a professional attitude is also a political attitude. Unless we take seriously the politics of our work, we will reduce ourselves to blind, sleepwalking experts in the hands of whoever wants to feed us.

Part IV

Practice

It is illuminating to see how our thinking about computers is rooted in a mechanistic world view, and how our work with information systems for organizations brings us up against romantic challenges. It is important to be aware of the role of systems thinking in our profession and to see that there are radically different ways of thinking about systems. We can increase our understanding of our professional role by taking part in discussions of the different paradigms for software development and the many aspects of the problem of quality. But all this illumination and understanding will come to nothing unless it is confronted with and integrated into our practice. Philosophical reflection is idle if we are unable to change our practice.

In the first two parts of this book we examined the subject of systems development by discussing its products – that is, systems – and by discussing the process of production – that is, development. In part III we used the notion of quality to discuss the raison d'être of our profession, and its fundamental ambitions and standards. Now, in part IV, we shall examine the essential contradictions involved in the everyday work of a computer professional.

In our practice we all have to deal with the contradictions between computers and people and between systems and change. In chapters 10 and 11 we try to make philosophy and practice meet by showing how philosophical theories can deepen our understanding of these contradictions. Our interest is not in analyzing the details of those theories but in illustrating the relevance of philosophical reflection to the very concrete problems of practice.

In the last chapter we summarize the many different perspectives introduced in the book, and we illustrate how to bring philosophy closer to practice by playing with metaphors and role models. We discuss how the conscious use of different perspectives can make us aware of our own presuppositions and prejudices. Without such an awareness, there is no hope at all for a deliberate change of practice. Philosophy can help us see what we do and why, but that is as far as it goes. People can change only in practice.

10

Computers and People

It is the painful experience of most people involved in systems development that personnel shortcomings – in qualifications, attitudes or experience – are a major source of errors and risks. But why is it, then, that the number one remedy suggested by the very same people is the application of new methods and tools?

All the way from machine code to assembly languages, high level languages, fourth generation languages, and further on to CASE tools, technological inventions have been introduced in order to solve our professional problems and help us meet requirements and develop better systems. The same holds for methods like structured programming, structured analysis and design, object-oriented approaches and prototyping, conceptual modeling, and requirements engineering.

Each of these inventions has no doubt made an important contribution to our profession. But we have to realize that, through the decades in which these new tools and methods were introduced, the software crisis did not disappear. On the contrary, the software crisis became more manifest and widespread, calling into question the credibility and integrity of our profession.

Gerald Weinberg reminds us of the way we tend to handle problems involving people: "Some years ago, I decided to shift my attention from the technical problems of software engineering to the problems of people who lead software engineering projects. The shift has not been easy.... In shifting my work from technical issues to leadership issues, I had to give up easy technical success for practically insoluble problems with people. When I became frustrated, I started treating people like machines. It didn't work. Is it possible that some of our software engineering failures result from *trying to manage people as if they were computers?* After all, isn't that what we know best?"

Weinberg is saying that systems developers naturally tend to view the world from a computer perspective. When they get promoted from programming to management, they must use the knowledge they have in solving the problems they face. People are then seen as modules or program components and are treated by the very techniques used in the design of computer systems. The competence that served the systems developers so well when they were constructing systems, and that was the reason for their promotion, now becomes a burden rather than an asset in their efforts to deal with people.

The Computer Perspective

Our profession is preoccupied with computers and technology, with techniques and general methods. We are experts at finding technical solutions and our research and development activities have been influenced primarily by engineering and science institutions. We understand and restrict our profession in such a way that independently of what the problem or challenge is, we are most likely to come up with a new technical device for a solution.

To whatever extent we, in our work, pay any attention at all to people, we try to deal with them with the very methods we use with our machines. What else can we do? From our computer perspective, the challenge presented by people is not to solve their concrete problems but rather to develop technologies and methods that will make our dealings with people as professional as our dealings with the machines.

Our profession seems to be trapped in a paradox between the practical problems we experience and the innovations we produce to increase professional standards. This paradox is one of many expressions of a fundamental contradiction between humanism and technology, between people and computers.

Having a computer perspective can mean being blind to people, seeing only computers, thinking only of computers, and so on. But it can also mean seeing everything in its actual or possible relations to computer technology – that is, using computer technology to organize what we see. Having a computer perspective can also mean treating everything as a computer or computer system – that is, applying the principles that we judge computers by to everything we see.

Weinberg is mainly interested in the last sense, but the other senses are often more obviously operative in the everyday life of a systems developer. In the first sense, we don't really pay any notice to people; in the second sense, people are seen as programmers or users and nothing else; in the third sense, people are treated as information processors, components, or modules. Usually, these three versions of a computer perspective are mixed together into a powerful combination of ways of using the computer to organize our perception and thinking about technology and people.

With a computer perspective, people disappear into the background. But you can rest assured that they won't stay there forever. People have a tendency to make themselves seen, and heard, and often at the most inconvenient moments. Project managers trying to salvage a project in the aftermath of sloppy requirements work know this all too well.

Requirements definition is one of the crucial activities in systems development. On the borderline between analysis and design, the requirements specification is the result of an analysis of the existing information systems of the user organization, while being at the same time the first serious attempt to describe the new computer-based system. Requirements definition is a demanding task requiring understanding and insight as well as creativity and invention.

Users participating in requirements definition must have experience with the existing information system and organization. But they must also be able to abstract from this mode of operation and focus on essential properties and possible changes. Preferably, they should also know something about the type of technology involved in the specific systems development effort.

Participating systems developers must be able to extract from the existing system a logical equivalent describing the fundamental structures and properties of the information system apart from present technological and organizational choices. They must master the technology involved in designing the new system. And last, but not least, they must be able to establish and manage effective cooperation and communication with users and among themselves.

Shortcomings in any of these respects may easily turn out to be catastrophic for the project, and using even the best of methods will not change this. If there is not enough ability to abstract, the new system may turn out to be a simple reproduction of the existing system, or even worse, it may turn out to be ill-structured and difficult to maintain and modify. If there is no thorough appreciation of the new technology involved, the new system may not benefit from the potentials of this technology and the system may turn out to have unforeseen and negative consequences. If there is no ability to establish and manage cooperation, the project may be seriously delayed and the potentials of the individual participants may not be used constructively to produce a high quality result.

A computer perspective has its roots in the mechanistic world view that was first described in chapter 1. With the clockwork as exemplary mechanism and with Newtonian mechanics setting the standard for knowledge, the mechanistic way of thinking about the universe, society, and people had a robust foundation in machine technology and physics.

The foundations of computer technology, the fundamental ideas of all its supporting research and engineering disciplines, belong to this mechanistic view of the world. Just as Descartes and Newton were fascinated by the clockwork mechanism, we are beguiled by our machine, the computer. The computer is our bread and butter, our professional pride and prejudice, defining what we know and who we are. So, how could we seriously consider reckoning with people, except as extensions to computers, or as some (rather unreliable) sorts of computers? How could we even imagine a people perspective?

Having a people perspective would not mean thinking of computers as your friends, giving them names, talking to them, and so on. Perceiving computers as persons would rather be an expression of an extreme computer perspective. No, having a people perspective would mean either not seeing computers at all, or putting them in their place as mere tools, extensions of people.

People perspectives belong to romanticism, the world view introduced in chapter 2. The romantic poets and philosophers reacted against the mechanistic world view, introducing history as their new science. The romanticists had a people perspective and were mainly interested in culture as the more or less artistic expression of human creativity and the need for self-realization. When mechanism and romanticism clashed, as they did all throughout the nineteenth century, the clashes came in many varieties: reason versus emotion, science versus religion, technology versus people, and so on.

One of the more interesting such clashes concerned the nature and methods of knowledge, the opposition between natural science and the humanities. In this chapter, we shall study this clash in order to arrive at a deeper understanding of the contradiction between technology and people, between a computer perspective and a people perspective.

Out of the mechanistic world view grew a philosophy and methodology of science later to be called *positivism* by the French philosopher Auguste Comte (1798–1857). Most of the work at developing this theory of what scientific knowledge is and how it should be pursued was done already in Newton's time by the British empiricists. But Comte became famous for coining the term and for turning positivism more or less into a religion. Comte believed in science – that is, in the natural sciences and their methods – and he wanted to apply those methods in a study of society. It would take another 50 years or so, until the turn of the century, before we could speak of a social science, but at least in France this science owed a lot to Comte's influence.

The romantic world view grew out of an interest in art and history, already from the very beginning developing ideas about how to study art and history. These ideas were collected into a philosophy of interpretation, of texts, of works of art, and of history, later to be called *hermeneutics*.

As long as positivists kept to the natural sciences and hermeneutics stayed within the sphere of humanities, they could coexist in a sort of truce looking down on one another. But when either one tried to enter a territory claimed by the other, war broke out. One such battle was fought toward the end of the nineteenth century over the social sciences, and this war has been going on ever since.

If we look only at the history of modern science, it all begins with the natural sciences and positivism. Hermeneutics and an interest in the humanities come later, and partly as a reaction to a dominating mechanistic perspective. This order of events seems somehow natural, and we find it almost everywhere we look. Sooner or later in the history of a practice it will turn to science for advice, passing through a stage of positivism only to enter a more chaotic period of attempts to develop hermeneutic alternatives. We have seen this happen in education, management, medicine, and social work. And we are seeing it happen in systems development. We began by taking an interest in computers, only later to realize that there were people involved too.

Science Makes Progress Possible

Positivism and hermeneutics are philosophies of science, but they are much more than that. They can easily be generalized to become theories of professional practices more generally, and they can serve as models for theorizing about a specific profession like systems development. This should not surprise us.

Coming out of two very complex world views, both positivism and hermeneutics carry a rich heritage. Habermas derives two of his knowledge interests, technology and understanding, from positivism and hermeneutics respectively, and the two theories therefore represent two different views, labor and language, on how society is organized and could be changed. And, being philosophies of science, they reflect upon an activity that during the last 50 years has served as the model for almost everything else in our culture.

For Comte, positivism comes as the crowning of our civilization. Having passed through the stages of magic and metaphysics, we have finally reached the positive stage where we refrain from naively populating the world with supernatural beings or speculating about hidden forces and realities behind what we can observe by sensory perception. The name positivism indicates both the positive, optimistic, nature of Comte's belief in science and his view of scientific knowledge as being concerned only with the positive in the sense of the given, what can be observed with our senses. We shall now briefly describe five of the pet ideas of positivism: objective observation, explanation and prediction, general knowledge, hypothesis testing, and physicalism.

Objective observation rather than speculative theorizing is the hallmark of science. The scientific attitude, according to positivism, is one of objective detachment. As a neutral observer of a chain of events you minimize the risk of seeing what you want to see, of being prejudiced by your interests. Science becomes possible when you can make a clear distinction between yourself and the object you are investigating.

As long as you experience yourself as one with the world, when you are engaged in it, your physics will be biased by paying particular attention to what is useful. Similarly, a science of the body will become possible only when you can view your body as an object, somehow detached from yourself. When studying society you will, as a positivist, take care not to get involved in your object of study. You will take pains to remain a neutral, unobserved observer in order to represent facts as they really are.

Explanation and prediction are the major aims of science, according to positivism. You explain an event by pointing to its cause, and you predict a future event by observing its cause. To be able to explain and predict, you must collect causal laws so that you know what causes go with what effects. But positivists are not really happy with the notion of a causal law, since all that can really be observed is that one event regularly follows upon another. Rather than speaking of causes and causal laws, positivists therefore prefer to speak of empirical generalizations or regularities.

Observable events are explained or predicted by being deduced from empirical generalizations. We make such generalizations all the time, even if we don't bother to formulate them explicitly. Think only of how a child goes around learning the effects of all kinds of actions. It is the aim of science to systematize such empirical generalizations, of attaining economy by seeking to reduce them to the most general ones. Science is systematized common sense.

Explanation must be objective, according to positivism, or else it does not deserve the name. You have to point to a cause that anyone in the right place can observe. You have to refer to an empirical regularity that anyone with the right equipment can check. This requirement makes perfect sense when we see what positivists want to use explanations for.

Explanations are not for personal satisfaction. They are for the survival of humanity. You have to learn empirical regularities in order to survive, in order to predict and avoid dangerous events. Explanation and prediction are only expressions of an underlying interest in controlling your environment. With a knowledge of empirical generalizations, you can control events

"Yes, with the amazing new 'knife,' you only have to wear the
SKIN of those dead animals."

by manipulating the causes of those events. Knowledge is power, the power to control.

Knowing by what means a certain epidemic is spread, you can use that knowledge to either spread that disease or stop it from spreading, provided the cause is objectively observable and possible to manipulate. Knowledge in itself is neither good nor bad, it all depends on how you use it.

Positivists are interested in *general knowledge*, in collecting empirical laws that can serve as future basis for action in all circumstances that are similar enough. Positivists do not care for the particular object under study, viewing this object instead as an instance, a test case for knowledge to be used elsewhere. What cannot be generalized, what will not add to the accumulating store of knowledge, is not worth knowing, according to positivism.

The natural sciences have been extremely successful in arriving at general truths about nature by using the method of analysis. Analyzing objects into their constituents has made it possible to substitute a relatively simple and standardized world for a world of confusing and changing variety.

Impressed by this success story, positivists regard analysis as the scientific method of choice. Analyzing the human mind into faculties and society into its constitutive individuals, positivists run the risk of concentrating on general truths about details at the expense of giving us an understanding of mind and society as wholes. Positivists want to analyze society into its most basic components, thus looking upon individual human beings as comparable to the atoms of physical theory. The varieties of social organizations can then be understood as different combinations of fundamentally similar individuals.

Inspired by the success of the natural sciences, positivists believe in science. Science makes progress, and with the aid of science society can make progress too. Social progress is best attained by letting science and the scientific method serve as a model for other human activities. With a scientific attitude toward education, politics, social work, planning, religion, and ethics – toward all human affairs – progress can be seen in these areas too.

Science is first and foremost a rational activity. In science we test our hypotheses. We do not remain content with conventional truths handed down to us. We subject them to serious scrutiny. It is all very well to reach agreement as the result of a debate with arguments and open communication, but unless we subject whatever we agree upon to empirical testing we are certainly less than rational.

This fairly simple idea of *hypothesis testing,* if taken seriously, has formidable consequences. To test an idea, we have to know what idea it is, and the idea has to be testable. To make sure we know what we are testing, we have to formulate our ideas as exactly as possible. Mathematics being our ideally exact language, positivists prefer to put ideas in a mathematical form. Ideas that cannot be formulated by the use of mathematics, properties that we don't know how to quantify, become suspect and treated as not really rational.

Judgments regarding art, religious beliefs, and moral sentiments fare badly when we stress the importance of empirically testing our ideas. Even when we appreciate the underlying idea of subjecting our ideas to testing, it will become irritating in the long run to constantly meet with the question "How do you measure that?" or in our own jargon "How do you implement that?"

With physics as our model science, we tend to become advocates of *physicalism,* believing that whatever is true can be formulated as a truth in physics, that whatever is real is physically real. We want to reduce every truth, in biology, psychology or sociology, to a statement in physics. We will think of biological organisms, minds and societies as complex organizations of physical entities. And even if we realize that it may be practically impossible to reduce all sciences to physics, we will dream of a unified science on the basis of physics.

Physics is expressed in the language of mathematics. The dream of a unified physical science is also the dream of mathematics as a universal language in which all truths can be formulated. The language of mathematics differs from ordinary natural languages in having explicitly specified rules of interpretation. There are no ambiguities, no room for disagreement about what a mathematical statement means.

A machine can work with mathematics. Throughout the history of positivism we see a fascination for the project of mechanizing science, the machine being understood as a symbol of objectivity. With computer technology this dream has come true. We now have an objective criterion of exactness: implementability. That the program is actually running is an infallible guarantee of success. And as in classical positivism, the computer positivist will worry little about what did not get into the program, what could not be implemented, as long as the program is running.

Together these ideas – of objective observation, explanation and prediction, general knowledge, hypothesis testing, physicalism – constitute positivism. In other respects, of course, the positivistic view of the scientific method has changed as science has changed.

Speaking of science, we often tend to think of it as basically one and the same sort of enterprise from the early seventeenth century until today. But that is too simplified a view. Early science was very much a matter of collecting facts about the world. British natural scientists collected fossils from all over the world, bringing them home to the British Museum. These scientists practiced, and their philosophers believed in, the method of induction, in collecting facts and generalizing from these facts.

In the early ninteenth century, science entered the laboratory, turned from the collecting of facts to experiments under controlled circumstances. Bringing into the laboratory representative items from nature, the scientists subjected these items to severe treatment. These scientists had ideas about the workings of natural processes that they wanted to test. They practiced what has come to be called the hypothetico-deductive method, the testing of hypotheses by deducing from them testable implications.

Late in the nineteenth century, the laboratories began to grow, turning into factories, in which huge experiments could be performed, and where it eventually became possible to study nature by creating it. Today elementary particles are created in cyclotrons, previously unseen chemical substances are synthesized, and new organisms are being created by genetic engineering. Science has finally merged with technology, using construction as its method.

With computer technology, this constructivism takes on a new shape. We no longer have to actually create nature. We can design it. By simulating a natural process on a computer and then varying the parameters, we can investigate not only actual reality but possible reality as well.

Through all these changes in the practice of natural science, some things remain unchanged and give to positivism its character. Nature is under examination, and it is a nature that is in our power, that is an object for our inquiries. Whether we collect it and take it home to look at or we cut it up and take it apart or electrify it or create it or simulate and design it, nature remains passive and submissive. It has no voice, makes no objections to our violations. A scientific education will foster an attitude colored by the submissiveness of nature. But when this attitude is turned upon people, they will react. Hermeneutics is a voicing of this reaction.

True Knowledge Is Personal

Hermeneutics is derived from the Greek word meaning interpretation or theory of interpretation. We find the word in the name of Hermes, the messenger god with wings on his sandals, who brought messages from the gods to people. The word appears again in the seventeenth century when there is a revival of attempts to interpret the Bible. Interpretations of texts in the hermeneutic tradition typically try to go beyond the text itself to the intention of the author, performing the sin of speculation Comte warned against.

In hermeneutics there is interest in *understanding* actions or their products, such as texts or works of art, not in explaining or predicting events. To understand an action, we must interpret it and we must determine what action it was and why it was performed, with what intention. To understand a text or a work of art, we must find its meaning, we must be able to interpret it.

If the action is of a type you see all the time, there is normally no problem in understanding it, provided it is performed under normal circumstances. Someone thanking you for dinner gives you no problem provided you have served him dinner. A more unusual action can still be fairly easily understood if you can see that it is directed to a definite goal, and it is a goal that makes sense for the person acting to pursue. But there are actions that are neither habitual nor instrumental, and they can be more difficult to understand. Emotional expressions are not that difficult provided the circumstances are normal, but in order to understand them you need to have experienced the emotions yourself and you need to be close to the person acting.

Moving thus from habitual actions to emotional expressions we realize that the understanding of actions can be more or less objective. Often we understand another person's action because we know that person well, or simply because we care for that person, or because we have had similar experiences, and so on.

In such cases it is often impossible to defend our interpretation by pointing to objective facts. We just know, and our understanding is subjective. It may be frustrating to be unable to convince a third person that our understanding is correct, but what we wish for then is not more objective facts but a more sensible person. And the aim of hermeneutics is not the accumulation of objectively useful knowledge but the edification of the individual. Hermeneutics can be turned into a methodology for a professional study of the humanities, but even so it will go on stressing this fundamentally subjective motivation behind all quest for knowledge. All true knowledge is edifying, and as such knowledge is always good.

Against the positivistic idea of a detached observer, hermeneutics want to stress the importance of *participation*. Only by participating in a social process will we be able to really understand what is going on. There is no way we can learn about what goes on in an organization unless we communicate with its members, preferably becoming a member of the organization ourselves.

Against the idea of detached observation, hermeneutics put forth their idea of dialectic interaction as a method of understanding. By observing from the outside, we will be able to scratch only the surface of social life. To penetrate beyond the surface, we will have to engage actively in the life of the organization, we will have to go native.

Rather than observing the organization from the outside in order to arrive at possible suggestions for organizational change, we will as hermeneutics work with the organization, trying out organizational changes together with its members in what is often called action research. The interventionist, and to some extent the evolutionist, apply these hermeneutic ideas.

Hermeneutics are interested in particular knowledge, applicable here and now. They concentrate on the particular object under study, having no ambition to obtain generalizable knowledge. Believing that every situation is *unique*, hermeneutics do not really believe in general knowledge, and complain about the necessarily superficial nature of all such knowledge.

Being interested in general truths, positivists have a tendency to neglect the exceptional and abnormal. In contrast, hermeneutics want to stress the importance of understanding the world in all its richness and variety, of appreciating the abnormal and exceptional as keys to an understanding of the normal. Only by having an understanding of both mind and society in their entirety will it be possible to understand their elements, hermeneutics claim. Hermeneutics want to begin with society, viewing individual human beings as constituted by the society to which they belong.

You cannot talk to nature, but you can talk to people. People are subjects, not objects. If you treat them as objects, you will miss what is most characteristic of them and you will antagonize them. People will only succumb to the methods of positivistic science if they are oppressed or don't give a damn.

If you are interested in people, in what they do and what they have achieved, in how they organize themselves and what they believe in, you will have to approach them as one subject talking to another. Certainly you can observe them, collect statistical facts about them, experiment upon them, genetically manipulate them, and simulate their behavior on your computer. But in doing all this you will learn only about their surface, and this surface will change, will remain out of your control, due to the fact that you are dealing with people.

To really learn something worthwhile about people, you have to enter into a dialectic interaction with them. And then you will learn something not only about them but about yourself as well. Think about what it is like to reminisce about shared experiences with a friend, or what it is like to evaluate with a colleague the shared experiences during a project. Talking about the past, the two of you will piece together a picture that would have been very difficult to produce singlehandedly. In a conversation like this, you don't really learn anything new. All you do is bring to memory what you already know.

As time goes by, we reconstruct our past to fit our present. Talking to a colleague in your project makes it possible for both of you to confront the past, correcting each other's reconstructions. Such a conversation – such a dialogue in which you confront your ideas with someone else's – is an example of what hermeneutics call *dialectics*. It is really very simple and commonplace, but dialectics takes time and can be very threatening if taken seriously. Nature seems so much safer. Certainly it may contradict your hypotheses, but it will not object when you decide to disregard its verdict. You are in control.

If you want to understand a text fully, you have to read and understand all the sentences it contains. But the understanding of each sentence presupposes an understanding of the text as a whole. Is this a law text or a love letter, is it a manual or a short story? Is it a communication system or an accounting system?

In order to understand the text as a whole, you have to understand each sentence, but in order to understand the sentence, you have to understand the text as a whole. Your reading of the text, provided it is unknown and difficult to understand, will be a moving back and forth between its details and the whole.

The reading of a text exemplifies *the hermeneutic circle*, the dialectic going back and forth between the parts and the whole. The circle is not the perfect metaphor since most hermeneutics claim that there is both progress and an end to the process of understanding. There are many other examples of this kind of circle, but let us just look at one.

When two persons are trying to get to know each other, there will be a dialectic interplay in which each of the two responds to what the other has just said or done. Many of our conversations are not like this. Often we are just waiting for a chance to say something, using whatever excuse we can find to change the topic. We have a stock of routine programs that we have tried before and that we trust will do the trick again. But when we engage in a dialogue, we will adjust our comments to each other. This may result in hilarious misunderstandings taking us off on strange tangents, but sooner or later all dialogues reach the truth, or so say hermeneutics.

You can read nature like an open book if you know mathematics, Galileo claimed. The truths are there to be seen. But that is not reading, according to hermeneutics. There are no truths to be simply seen in a text. To understand a text, you must interpret it. And all interpretation involves a dialectic interaction with another subject. Unless understanding is mutual, there is no real understanding. I can understand you only if you under stand me.

Positivists tend to see objects everywhere. Hermeneutics tend to see subjects. History, to take one example, which to positivists is a succession of physical events, is seen by hermeneutics as a dialogue, a dialectic interaction between ideas about society and the good life. To understand history, we have to understand this dialogue. We have to see how events enter into this ongoing dialogue, responding to the past, putting questions to the future. We are part of that process ourselves, and our attempts to interpret the past are themselves a part of the very dialogue that is history.

We cannot step out of history to study it from the outside. History is not an objective process in objective time. Historical events take place whenever they appear in the dialogue, and they may reappear again and again. As we go on explaining hermeneutics like this, things may seem to get stranger and stranger. But if we just think of what a dialogue between two subjects is like, we see where hermeneutics get their ideas about history, and perhaps we even see that those ideas may be quite fruitful as a complement to our more mundane, positivistic understanding of history.

To positivists, history is a process of progress. To hermeneutics, it is a dialogue that we must enter into in order to become fully human. Still, at the end of history is truth, as in every dialogue, provided that we can remember its course.

But there will be truth at the end only if we don't cheat. And some of the more interesting ideas about interpretation and understanding are to be found in what has been called the hermeneutics of suspicion. Working toward the same goal, sharing the same interests, we will reach the truth. But when our goals and interests differ, we will disguise ourselves and try to conceal our motives. The dialogue turns into a power struggle. We shall see, in the next chapter, how very different our ideas about interpretation will be when communication is seen as a dialectic struggle for power rather than as a dialogue.

Hermeneutics – with their ideas about interpretation, about subjective understanding, participation, the uniqueness of situations, dialectics, and the hermeneutic circle – have a powerful alternative to the positivistic view of knowledge.

Against the positivistic method of detached observation, hermeneutics pose their idea of gaining knowledge by subjective participation. For positivism, the scientist is an outsider, an expert who collects information about people, if possible without revealing that they are subject to investigation and definitely not revealing for what purpose the investigation is taking place. The positivistic scientist is really spying on people in order to secretly learn the unbiased truth about them.

For hermeneutics, everyone is a scientist if anyone is. Information about an organization is gained within the organization and shared by all its members in an ongoing process of communication. Knowledge is not a commodity to be collected under controlled conditions, to be bought and sold on a market; knowledge is subjective enlightenment and edification, difficult but possible to share, provided there is mutual respect and a sincere attempt at understanding.

In Search of Excellence

We try to manage people as if they were computers, says Weinberg, because computers are what we know best. The competence we have, our professional expertise, is mechanistic through and through. In the short sketch of positivism given above, we find many examples of the kinds of things we know and value.

As systems constructors, our work begins with a well-specified data processing problem. That is our given. The challenge is to construct a system that meets these requirements and then, preferably, to prove by scientific means that the requirements are met. Using mathematical proofs to ensure quality, we are not questioning whether the requirements really correspond to what is needed, or whether the specification of the computer system is a complete description of all its qualities. As systems constructors, we do not go beyond the requirements to speculate about user needs or problematic situations in the user organization.

In systems development we are naturally concerned with empirical generalizations and regularities. We say "naturally" because using the computer requires formalizations, at least on the level of expression and in most cases on the level of content as well. If we want to use computers, we have to abstract from variations and concentrate on regularities in the way things are done. The computer invites us, and to some extent forces us, into a positivistic approach.

We can go on like this, but it is perhaps more interesting to stand back a little and look instead at even more fundamental properties of the positivistic view of knowledge.

Positivists are obsessed with control. How can we know anything unless we know that we know – that is, unless we have control of the process of obtaining knowledge? The notion of control defines both the means and aim of science. The scientific method teaches us how to control our experiments, how to use control groups, how to be objective, how to define our terms, how to use exact, quantitative measures and formalized language, how to reason, how to make our ideas clear and explicit, how to say what we mean and mean what we say.

The aim of knowledge, says positivism, is explanation and prediction. Explanation and prediction are ingredients in control. If we can explain our mistakes, we can avoid them. If we can predict an event, we can control it, provided we can manipulate its cause. Knowledge is useful. Knowledge is power. With knowledge we can gain freedom by learning to control whatever it is that is now controlling us.

The taxonomy for maturity levels in software development suggested by Watts Humphrey is a powerful attempt to apply this positivistic idea of control to our search for excellence in systems development. As discussed in chapter 8, Humphrey distinguishes between five levels of software-process maturity – initial, repeatable, defined, managed, and optimizing – five levels of what is obviously increasing control over the production process. To go from the initial to the repeatable level, we have to uncover patterns of activities to be reapplied again and again in different projects, and we have to implement basic management control to ensure that these patterns are, in fact, adhered to. To increase our control, we have then to explicitly define these patterns of activities, thereby creating a unified framework for process measurement within and among projects. Having thus reached the defined level of maturity, it is possible to develop the organization further. If we successfully reach the optimizing level of maturity, this means that we are able to effectively predict and control our systems development activities.

Positivists collect knowledge. Whether it is facts, or the generalizations based on facts, knowledge is an objective commodity that can be measured, bought, and classified. Knowledge is rather like information. It can be stored and transmitted, lost and retrieved; it can be stored in books, on hard disks, and in people. Teaching is the transmission of knowledge from one storage place to another.

Requirements specifications and systems development methods are expressions of this positivistic notion of knowledge. Requirements specifications are the result of collecting facts about present activities and needs in combination with facts about technological possibilities. Often, these requirements constitute the key part of a contract to be used to control and evaluate the development effort and the resulting computer system. Systems development methods are generalizations based on previous development efforts and theoretical constructs. They are documented in books, supported by various tools, and distributed from the authors to be used by a great number of systems developers in different organizations.

But knowledge is not just any commodity. To positivists, there is nothing as precious as knowledge. Organizations make great efforts to use new computer technologies, even if it is questionable whether such technologies are economically feasible and functionally necessary. Likewise, systems developers make great efforts to learn new programming languages and methods. They participate in expensive courses and seminars where gurus of our field preach the latest fashion. To be in control, to be in the know, to be an expert in the field, to know more than others – that is what excellence is all about.

The hermeneutic view of knowledge, its goals and means, is obviously very different. The aim of knowledge is enlightenment and edification. Knowledge is not a commodity on a market but a property of persons. Its value lies not in its power and price, but in the way it enriches our lives and contributes to our personal growth.

This means, of course, that it is not so much the quantity of knowledge we have as its quality that counts. It is not the number of facts we know that is important, but our understanding of the world. Teaching is not the transmission of information but the cultivation of the student. Important aspects of knowledge cannot be made explicit. They remain tacit and, like taste, they have to be acquired rather than collected.

Excellence lies not in the amount of knowledge we have but in the way we are. While our authority as experts rests on having all the facts and figures right, our authority as people lies in the way we take command without the use of facts and figures.

This hermeneutic view of knowledge is fundamentally antithetic to the mechanistic view dominating the computer professions. It is different enough to seem utterly irrelevant to our business. Not only that, the hermeneutic view seems so unscientific and steeped in mysticism as to be difficult to take seriously, not to say understand.

But we are not saying that we have to turn to hermeneutics to find out what people are really like. Instead, we are describing the hermeneutic view of knowledge in order to show, as a contrast to our own computer perspective, what a very influential people perspective looks like. To those that might argue that the hermeneutic view is too extreme, our answer would be that our own computer perspective is no less extreme.

The obvious remedy to Weinberg's complaint would be a crash course in social psychology or organization theory. If the reason we treat people like machines is that we know about computers rather than people, then let us have a few courses on people. This is fine, but we think there is a deeper aspect to Weinberg's complaint.

It is not just what we know but the way we organize what we know, our attitude toward people and to knowledge that is important. And hermeneutics is interesting not because it fills in the gaps in our knowledge but because it presents us with a different perspective on technology and people. A crash course on people that did not question our mechanistic world view would do very little in the way of changing our tendency to treat people like machines. It would simply be a course on those wonderful, but strange, machines called people.

By looking closer at the hermeneutic perspective, it is possible to see more clearly the role of people in the design and use of computer technology. For example, hermeneutics can help us see what is really going on when we use experiments and reviews as quality assurance techniques. We may think that we can use these techniques to simply determine the quality of software, with some uncertainty to be sure. But hermeneutics will help us understand the complex interplay of people, methods, and technology, and the important role of interpretation, personal interests, and values involved in the use of these techniques.

Hermeneutics will make us pay more attention to the complexities of the task of requirements definition. In this work as systems developers we have to engage in dialogues with users and clients. We have to interpret their professional languages to appreciate both the routines and irregularities involved in performing the relevant tasks in the user organization.

From a hermeneutic perspective, requirements specifications and systems development methods will not seem very important. It is the process of creating the requirements specification and the process of interpreting and changing it that is important. The quality of the process depends on how well we understand the present procedures, the needs and ideas of the users, and the ways in which computer technology can be used to improve the situation. It also depends on how well the users see and understand the process and on how well we communicate and interact.

Likewise, systems development methods are of little practical importance. It is our personal understanding and insights that determine the quality of our work. We cannot improve our competence simply by studying new methods, reading books, and attending professional seminars.

The struggle between positivism and hermeneutics will not be settled. It expresses the fundamental and antagonistic contradiction between technology and people. This contradiction is inherent in all our activities and ideas in systems development. Understanding and respecting the dialectic relation between computers and people, and more generally between technology and people, is one of the richest sources for developing a professional attitude toward the development of computer systems in organizations.

11

Systems and Change

Process and Structure
The Power of People
The Inertia of Culture
Designing for Change

Should a country be governed by rules that cannot be changed even if the majority of people and the elected parliament want to change them? Or should it always be possible for people to adjust their system of government to a changing environment? Certainly, it is the people who should decide what is right and what is wrong. But then, is it not undemocratic to permit a majority to deny civil rights to some citizens, or to restrict the freedom of speech for everyone? Should a country have an unchangeable constitution, as in the United States, or should it not, as in the United Kingdom? Which is the best way to safeguard democracy?

These are difficult questions, and the most sensible response is probably to suggest some sort of compromise. But it is not easy to specify the details of such a compromise. And what should be the status of that compromise? Should it be possible for the majority of the people to change it?

Systems development projects are compromises, managed both by rules and by influence of a changing environment. Some management initiatives are rule driven and others are situation driven.

The primary concern in rule-driven management is to structure the process by using project plans and organizational arrangements suited for the type of task at hand. Based on structural descriptions of how a project like this should proceed, the rule-driven manager monitors the process and intervenes to keep it on its predetermined course.

In contrast the primary concern in situation-driven management is to appreciate the uniqueness of the situations in the project. The situation-driven manager responds to new options and challenges as they appear, and intervenes to facilitate further development and change. Project plans are, of course, required, but they are only a secondary concern. The situation-driven manager is studying the process and uses project plans as an instrument of observation.

No project is managed in a purely rule-driven or situation-driven fashion. Project managers know that all projects require a mixture of both philosophies. The challenge is to know how to balance the two and how to adjust the balance when needed.

Some general advice can be given: the earlier in the project, the more likely is the need for a situation-driven approach; the closer the deadline, the stronger is the need for a rule-driven approach; and the higher the uncertainty of the task, the less applicable is the rule-driven approach. In practice, such general advice does not provide much support. The appropriate balance will always depend on the situation and the actors involved.

A similar dichotomy applies to the systems we develop. It must be possible to adapt the systems to deal with new situations in a changing environment, but the systems must also support us in handling every situation in accordance with established rules and procedures.

Modularization, encapsulation, and information hiding are all techniques designed to ensure robustness while at the same time allowing for flexibility. The idea is to let the fundamental structure of a system reflect the stable properties of the environment, and then to identify and hide the volatile properties within this structure. The volatile properties are encapsulated in such a way that changes affect the system only locally.

If changes occur in the environment, implying that the system must change, we have to modify only local details, whereas the fundamental structure can be preserved. This is true, of course, only to the extent that we have correctly identified the properties of the environment that will prove to be stable over time.

The best we can hope for is to design systems such that *given* a prediction about the environment of the system, modifications involving the volatile properties will be relatively easy to perform. There is always a risk that there will be other changes in the environment requiring dramatic and expensive changes in the fundamental structure of the system.

Process and Structure

Flexibility is, or should be, one of the most controversial issues in systems development. As systems designers and project managers, we design structures, which for better and for worse set the conditions for the daily work of users and systems developers. In that sense our discipline is essentially defined by a bureaucratic, rule-driven philosophy.

Structural arrangements will, however, sooner or later become burdens and obstacles to change. Changes in the environment will, eventually, force the fundamental structures of computer systems to change. Likewise, systems developers face unforeseen situations and they learn as the process develops, forcing project managers to modify or change existing plans and organizational arrangements.

In contradiction to its essentially bureaucratic nature, our discipline is facing a fundamental requirement to deal deftly with change. Computer systems are rarely introduced in order to support a traditional way of operation. Instead, investments in computer technology are made to change traditional ways of doing things, in order to solve problems and improve efficiency. How do we cope with this contradiction between stability and change?

We call what we do systems development, indicating that our task is development, and that what we develop are systems.

We develop – that is, we are engaged in efforts of creation, taking part in ad hoc projects characterized by the turbulent chaos of innovation. But what we develop are systems, stable structures providing an order difficult to change and deviate from. Living in the midst of change, we must create something stable. Does this make sense?

We talk of processes to indicate phenomena of movement or change. In contrast, we say that properties that are fixed and stable are structural, or belong to the structure. A program is or has a structure. A program execution is a process.

When we get interested in a software house, we can decide to focus on its more stable properties, on its structures. We will study the architecture of its buildings, its technology, employees, products, services, and so on. But we will also look closer at its procedures and methods, the division of labor, coordination mechanisms, departmental divisions, and so on.

Alternatively, we can focus instead on the way people do their jobs, the way products and services are produced, and the way some people manage other people. We can study the activities aimed at changing the way the software house operates: professional training, technological development, new products and services, as well as informal change processes. Taking such a view, we will learn about organizational processes rather than structures in the software house, focusing either on the daily routine or on organizational change and development.

In this way we can use the concepts of process and structure to characterize two complementary aspects of a software house. This is all very simple, at least as long as we attend to one or the other of these aspects. As soon as we get interested in how processes and structures are *related*, however, we will begin to appreciate the deep contradictions between the two.

Beginning from the structural end of the software house, we can identify two different relations to processes. First, the software house is a structural arrangement controlling the daily routine of systems developers, administrators, and managers. The software house is a system for producing specific products and services. Second, those structural arrangements change as a result of organizational change efforts. CASE tools are introduced. Old departments disappear and new ones are created.

Whereas the first relation is as stable and harmonious as the daily routine, the second relation is contradictory in nature. Professional training may not have the intended effect because it fails in altering well-established and structurally supported work habits and beliefs. New CASE tools might not be used effectively because systems developers continue to rely on old routines and methods. Just because structures serve as stable frameworks for processes, they will protect those processes against change. Since our interest in change is an interest in processes, in how things are done, we often try to change processes directly, only to find out how difficult or even impossible it can be to change a process without changing its supporting structure.

Beginning instead from the process end, we can focus on how activities, in an evolutionary, harmonious way, change structural aspects of products, services, and organizational arrangements. We focus on the work being done in software projects and on how this gradually changes different structures. But project groups perform their task in a potentially contradictory relationship with their contractual arrangements. As the project develops, new insights are gained and unforeseen situations emerge. Suddenly, the process may change orientation. The primary task of the project slides into the background as project members become more concerned with changing the structural arrangements governing the project. Ad hoc meetings are held, negotiations take place, and as a result new contractual arrangements are established.

The introduction of a new computer system can amount to a major structural change in an organization, deeply affecting management structures, the division of labor, and daily routines. To the software house installing the system, it can very well be just one of many similar projects, part of an established and rather boring routine.

Examples like these suggest a rather simple but magnificent view of systems development, inviting us to see ourselves as change agents, responsible for constructing and delivering systems for others to use. But systems developers are obviously not the only ones developing systems while everyone else is simply using them.

Users perform their tasks as part of a daily routine, and computer systems are part of the structural arrangements governing this activity. But the environment changes, users improve their skills and create new visions about effective ways to use computers. Similarly, systems developers perform their tasks as part of their daily routine. They use computers, projects plans, and organizational arrangements to structure their activities. But requirements change, new solutions are suggested, and unforeseen problems emerge.

Is there really any difference between the two? Users are certainly not just performing their tasks in a harmonious way within stable structures, as parts of a system. And systems developers are not all the time engaged in change processes, breaking down existing structures and creating new worlds.

The Power of People

In chapter 10 we drew on two theories of science – positivism and hermeneutics – to reflect upon two perspectives – computers and people. We can use two other theories – structuralism and critical theory – to deepen our understanding of the relations between structures and processes, between systems and change. Positivism is the view of science dominating a mechanistic world view. Hermeneutics is at the heart of the romantic revolt against this world view. Both *structuralism* and *critical theory* belong within a romantic world view, but they developed as reactions to central hermeneutic ideas.

Positivists with their stress on objective knowledge, with their scientific attitude of objectifying phenomena, can deepen our understanding of a perspective on computer use focusing on computer technology. Hermeneutics with their interest in subjects give us a richer foundation on which to form a people perspective on the development and use of computers. Structuralists will argue against hermeneutics in favor of a more objective view of social phenomena, but they will stress, contrary to positivism, the important role of interpretation. Critical theorists share with hermeneutics an interest in historical processes, but they have a very different understanding of those processes, seeing in them conflict rather than harmony, contradictions and power struggles rather than earnest attempts at mutual understanding.

Classifying ideas into a few categories like this is, of course, to simplify and distort a confusing variety of different ideas and thinkers. But our interest here lies not in intellectual history. We want to obtain a clearer understanding of different perspectives within systems development.

Critical theory began, with the establishment of the Institute for Social Research in Frankfurt, Germany, in the early 1930s as a Marxist alternative to a positivistic social science. Marxism had by then begun to harden into the ideology of a dominating Soviet Communist Party, which had spawned subservient parties all over the Western world.

The critical theorists were not particularly interested in the Marxism preached by those parties. Rather than reading Marx as an economist, the researchers at the Frankfurt school studied his early, more philosophical works. In those at that time still unpublished manuscripts was found a different Marx, more humanistic than the one portrayed by Lenin and Stalin. Two of the young Marx's theories were to become particularly influential: the theory of alienation and the theory of historical materialism.

Developing a basically romantic view of the individual, the young Marx had stressed the importance of self-realization through expression. A painter identifies with her paintings. Pointing at them, she can say "This is me." These concrete expressions of herself together constitute her self. She becomes real to others and to herself through these concrete, objective expressions. When others take an interest in her paintings, they take an interest in her, they acknowledge her. When they are uninterested, when they don't even look at her paintings, they reduce her, make her doubtful of what she has done, doubtful of who she is, doubtful if she really is anything.

When others acknowledge us by attending to and commenting upon our expressions, attributing them to us, we become self-certain, certain in our selves. When they don't take notice of our expressions, these expressions will seem less real, and we will become uncertain in our selves. When they attend to our expressions but refuse to attribute them to us, when they systematically attribute them to someone else, we lose control over who we are, and we become aliens to ourselves.

This theory of self-realization through expression is not a bad theory of what it is like to seek one's identity through creative work. But according to Marx this is what we all do, all the time. As little children, we express ourselves, hungry for acknowledgment, and we develop our identities as a result of the acknowledgment we get. As adults, we express ourselves, not only in paintings but in the families we raise, the work we do, the homes we decorate, the cars we drive, the clothes we wear, the food we cook, the plants we grow. As systems developers, we express ourselves in the specifications, prototypes, and systems we produce and also in the project groups we belong to and the committees of which we are members.

Wondering who we are, we point to all these things, and others help us in affirming our identity. But, and this is the thrust of Marx's argument, modern industrial production is organized in such a way as to minimize the possibilities for self-expression. Shoemakers used to make shoes on order, from beginning to end. Modern societies produce shoes on production lines for the market, each worker contributing a standardized operation somewhere along the line. A robot can do it, and indeed often does, the worker being reduced to taking care of the machine.

All day long the worker is engaged in producing shoes, but it is not the worker who makes the shoes. In no way can he point to the shoes in a shop window saying "That is me." So, who is he then? How does he realize his self? Well, he has to do it in his leisure time, when he is off from work. For the majority of people, work is no longer a means of self-expression. It is only a way of making a living, a living that takes place after work.

The worker in an industrialized society is alienated. Since what he produces is in no way his expression, who is he at work? His fellow workers will not know what or who he is, and he will be an alien to them. Since what he produces is not his, he will be alienated from his self-expression, an alien to himself.

This alienation at work can be set right only by changing the system of production. As long as others own your work and thus control your major means of self-expression, you will remain alienated. But work enables you to make money. The industrial production system is economically sound and it gives the workers a lot to spend. And as long as they spend it, consuming all the gadgets being produced, the wheels of production will turn, and there will be more money to spend.

But spending money is not a particularly interesting means of self-realization. The life of a consumer is not much to look back upon. Mass consumption makes for mass people, for one-dimensional men who are what they consume, and in the consumer society alienation spreads.

Computer technology is used, in line with other technologies, to industrialize work. It therefore makes sense to discuss the alienation of the users of that technology. But what about ourselves? What is really the effect of using structured techniques, formal reviews, chief programmer teams, and CASE tools in systems development?

The very idea of egoless programming is to make systems developers feel less personally involved in the products they produce in order to allow others to criticize and improve them. And many of the techniques and methods we apply are simply organizational arrangements to facilitate division of labor and to improve efficiency.

Normally, we think of systems development as a challenging and creative profession. But systems development is, of course, just one of many professions in our industrialized societies. Certainly, it makes sense to ask whether systems developers are alienated in their work.

By combining Marx's theory of alienation with a theory of rationality, the Frankfurt school launched a powerful attack on our modern high-tech society. To be rational is to choose, to the best of our knowledge, the means best suited for whatever aim we have in mind. Rationality is instrumental, or at least it is so in our modern society.

The process of modernization, beginning already in the sixteenth century, was conceived of as a process substituting rational ways of doing things for traditional habits. Science was substituted for magic, a rational bureaucracy with explicit rules replaced a corrupt administration dominated by nepotism, democracy gradually undermined a society based on family privileges, and industrial production with machines took over from labor-intensive craft.

All these areas exemplify slightly different ideas about rationalization, but thanks to the dominating position of industrial production in our modern society it was rationality in the sense of efficiency that eventually came to dominate the others.

To be efficient, to economize with resources, to maximize profit, became the standard by which we moderns judge ourselves and others. We admire efficiency without being too concerned with goals or with the quality of results. Efficient dishwashers, efficient surgeons, efficient killers, efficient painters, efficient quarterbacks – whatever the job, we admire it if it is done fast, smoothly, and with a minimum of effort.

Of course, we all know that it is less than rational to work efficiently toward a silly goal. "If something isn't worth doing, it isn't worth doing well." But who is to say which goals are worth pursuing? And by what standards are we to evaluate the quality of our results? If I want something, who is to say I should not want it? By making instrumental, technical rationality our fundamental value, we want to say that only the means can be objectively judged. The goals and the quality of the results are subjective, beyond dispute and criticism.

Such a value relativism, or really such a negation of values, is difficult to accept in practice. A society needs something more definite in the way of values in order to function, needs commitment to common goals as well as to efficiency. And those common goals have to be discussed and developed if we want to pride ourselves on being rational.

But what are those goals? They have to do with what we consider the good life to be, with what we think is worth doing, with what we believe it means to be a human being, a person. If we are all alienated, caught in a rat race of producing more and consuming more, our only idea about the good life is "more." We are one-dimensional and certainly less than rational, in spite of our admirable efficiency. Our only chance, says critical theory, lies in curing ourselves of our fascination for technical rationality to become able to think more clearly about what is important and what is not.

Why are we so taken in by efficiency? How can a quality of good technology turn into a human value? Marx's answer to these questions is his theory of historical materialism.

Materialism, in this sense, is the theory that the character of a society is determined by its mode of production. The fundamental human condition, according to a materialist, is labor. We have to work to live. It is because we have to work that there are societies in the first place. A society is basically a mode of organizing work, a mode of production. Everything else – politics, laws, schools, families, science, entertainment, religion – is part of that organization and therefore determined by the way labor is organized.

According to historical materialism, not only is society materially based, it is also an ongoing process of change, where lasting changes always reflect changes in the mode of production. Since technology plays such an important role in modern modes of production, this theory of social change seems to be an example of technological determinism. Most Marxists, including Marx himself, and especially the critical theorists, are unhappy with this reading of historical materialism, however.

Critical theorists share with hermeneutics a belief in the power of people to create their history. Combining this with historical materialism is not easy. You can try to argue that even if everything depends on the mode of production, people can change society by changing those modes of production. But they have to do so in a manner not determined by the modes of production themselves. If we are just puppets, determined in our thinking by the material basis of our society, the changes we will bring to that basis will not originate in us but in the material basis itself.

It is disputable whether it is possible to combine a belief in the power of the subject with historical materialism, but that is what critical theorists want to do. They see as their major task, and this is the major task of a critical social science, the liberation of people from the power of technological determinism. By showing how ideas that dominate our society are determined by our mode of production, they want us to question those ideas. Why is it, for example, that atomism seems such a powerful idea in modern science? Well, look at the atomism in our ways of organizing work. Is not the production line and the whole idea of the division of labor a powerful example of atomism? No wonder we think it is a good idea, when our society is founded on it.

The Inertia of Culture

To critical theorists, society is first and foremost an ongoing process of change. To structuralists, any society or organization is a stable structure, and an understanding of social phenomena is possible only by identifying and analyzing them as structures. But structuralism is really a theory of language, developed in the early decades of this century as a reaction to a dominating hermeneutic interest in the history of languages.

The hermeneutics were fascinated by the problem of tracing the different European languages back to their Indian sources. Believing language to be the most important element of culture, the hermeneutics argued that people were shaped by their language. In language live the traces of a long-forgotten past. Learning our language, how its words were once formed and the changes they have gone through, we learn our history. Language has been shaped and reshaped by tradition. Thus the collected wisdom of our ancestors lies buried in our language, for us to interpret and decipher. This interest in the historical roots of language led to an interest in the etymology of individual words and the ancestry of language families at the expense of an interest in language as a complex rule-governed system.

Structuralists argue, against this, that language itself, rather than its history, should be the focus for a romantic theory of language. Languages, be it programming languages, professional languages, or natural language, are systems of signs. Signs get their meaning from their relations – which may be opposition, difference, or similarity – to other signs in the system. The relations between signs together form a structure. The meaning of a sign is its place in that structure. To determine what a color word means, you have to find all the color words in the language and see how they are related to one another. If there are only two color words, as there are in many languages, it does not make sense to give them a meaning richer than the one possible with such a simple structure. The interpretation of a sign is always relative to what other signs there are in the system.

Structuralism began as a theory of language, and it would have remained interesting only to linguists if it had not been generalized into a theory of all kinds of systems of signs. We have already seen an example of such a generalization in the structural analysis given by Lévi-Strauss of the architecture of the Bororo village.

Lévi-Strauss wants us to see that the Bororo architecture, or village plan, is a means of communicating the culture of the Bororo, each hut being a sign in a complex system. Myths, folk tales, advertisements, family structures, movies, meals, and sports events are only a few examples of phenomena analyzed by structuralists as complex systems of signs. Thus structuralism is generalized into a broad approach to the study of social phenomena and human activity, stressing the communicative aspect of such phenomena. In such a structural analysis we are not particularly interested in the history of the phenomenon. We want to stress instead the stable, systematic, rule-governed nature of such systems of signs, wanting to interpret a particular sign by placing it in the system to which it belongs.

Often we are interested in seeing how similar structures appear in different settings. One of the major inspirations for structuralism has been a now classical analysis of Russian folk tales, showing how they all exemplify the very same structure: the roles are the same, the basic actions are the same, the means used are the same, and so on. Every tale really tells the same story.

We have already given an example of such a retelling of one and the same story in Michel Foucault's analysis of the major institutions of modern society as communicating a single message of control. If you are a materialist, you will want to show, of course, that the basic structure to be found in the way labor is organized will recur again and again. In medieval Europe, labor was hierarchically organized into a great number of levels from king to serf. No wonder there was such avid interest in organizing both heaven and hell in likewise fashion: god, archangels, angels, saints, and devil, archdevils, devils, and so on with numerous grades in between.

Lévi-Strauss turned structural linguistics into a general theory of how to study social phenomena and human activity. His study of illiterate cultures colored his version of structuralism.

Early in his career, he was struck by the stability and complexity of such stone age societies. For several thousand years, these societies remained virtually unchanged, he argued. Their population did not grow or diminish, their habits did not change, and their language, the myths they told, and the tools they used in their work remained the same.

Being impressed by these societies, Lévi-Strauss began to view social change as an abnormal phenomenon. What we call modern civilization is really an aberration. The whole point of social institutions, of culture, is to secure the stability of social life, to organize social interaction. To this end, a society or an organization will use a number of different means – the architecture of villages and buildings, kinship systems and employment regulations, myths and legends, names and totems, and rules for dressing, cooking, eating, working, and so on – all being systems of signs communicating a certain way of life, a certain way of organizing social interaction.

A person growing up in such a culture will internalize its structures and will be shaped by them. Indeed, were it not for our cultures, be it on the level of societies or on the level of organizations, we human beings would just be another type of animal. We are like empty containers to be filled by culture, acquiring our languages, ideas, habits, and dreams from culture.

A culture, understood in this way as an interrelated set of systems of signs, really does not need people to exist. It needs people to change – for in the transference from generation to generation cultures will change – but if there is no desire to change, people are not really important. Thus Lévi-Strauss follows his mentors in French sociology in favoring an extreme version of social holism, of stressing the dominance of social structures over individual subjects. We are nothing but puppets in the strings of our society or culture, caught in the structure like small fish in a huge net.

From this point of view, the very idea of systems development seems extremely ambitious. Most user organizations believe that they can change and improve their operation by developing new computer systems. And most systems development organizations invest money, energy and prestige in improving productivity and quality by introducing new methods and computerized tools. Are we, maybe, driven too much by utopian dreams? Are we naive in not understanding and respecting the practical inertia involved in social change?

A structuralist like Lévi-Strauss is not interested in change but in the stability of society and organizations. It is not the process but the structure that interests him. Still, Lévi-Strauss is a dialectician like all other romantic thinkers. The basic building block of all cultures, of any structure, is the binary pair of opposites, the contradiction. As Lévi-Strauss tours the world of cultures, he is struck by the universality of dichotomic thinking. Every structural analysis will proceed on the hypothesis that all systems of signs are organized as oppositions between contradictory ideas. But these contradictions are not grounds for change, as they so often are in romantic thinking.

The fundamental contradiction is the one of nature versus culture. A human being is an animal that becomes filled with culture. This contradiction between nature and culture is not one that can be resolved, but one that we have to live with. Everywhere in every culture this fundamental contradiction reappears. Indeed, a culture that fails to balance this and other basic contradictions of human life – such as that between life and death – will have to change, or go under.

More generally, Lévi-Strauss' attitude toward the contradictions of a society is to look for the ways these contradictions are stabilized, how they are handled so as to permit the society to live with them without changing. Myths, religions, and ideologies serve as correctives when the contradictions in everyday life are only too obvious. Why we have to die will be explained to us in a tale of original sin, why we have to work is likewise explained. Why some have to starve while others eat themselves to death is explained by telling stories about the difference between races or the necessity of a free market.

Modernization has meant a demystification of the world. We no longer tell myths to one another, and the doctrine of original sin does not help us much in coming to terms with death and work. But to think we can do without storytelling is naive, and when we look at the stories we tell in soap operas or in science, the differences between them and the old myths are not particularly striking. And to Lévi-Strauss it is pretty much one and the same thing. Like most romantics, he does not think of science, and he would not think of systems development, as particularly rational.

Designing for Change

Because we are interested in what we can learn from these theories of science, let us see what consequences we can draw from them about the task of systems development, the task of constructing or designing computer technology for human use.

Positivists have a straightforward view of systems development as engineering, as putting to human use our knowledge of causal regularities in material and social reality. The ideal of systems construction is based on positivist ideas. Engineers tend to focus on technical issues, or they consider physical and social engineering as fundamentally similar. The ambition is always to construct systems, physical or social, with a maximum of efficiency and control.

Hermeneutics stress instead the role of understanding in the construction of reality. Systems evolution as a strategy and the role of the facilitator both rely on hermeneutic ideas. Reality becomes real only through our understanding of it. What we do not understand does not really exist. Technology is what we understand it to be. Only to that extent will we use it and experience it as useful.

Critical theorists are engaged in projects of deconstruction of the materially based ideologies that prevent us from developing our interests as human beings. The strategy of intervention and the role of the emancipator are strongly inspired by critical theory. The early critical theorists turned to art in an attempt to emancipate us from the grip of technology in modern industrialized society. As systems developers in the role of emancipators, we have a more optimistic attitude toward technology. We want to see systems development as both a process of critical deconstruction and a constructive process, using computers not only as means in doing our job but as an opportunity to emancipate ourselves from technological and organizational alienation.

Structuralism, as a theory of science, is more interested in analysis than in construction. Analyzing technology in its use as a system of signs, structuralists want us to see what the technology really means, what it communicates to the users. But we can turn the guidelines of structural analysis into a theory of design, stressing the importance of the symbolic qualities of technology, in addition to their functional and aesthetic ones.

Such a design theory would stress the structural properties of computer technology, how it takes its place in already existing structural arrangements and how it contributes to and changes those arrangements. Computer technology invites us to attend to and describe structural properties, both on the level of individual computer systems and on the level of networks and information infrastructures. To this interest in structural relations or systems, a structuralist theory of design will add an appreciation of the role of these systems in the life of the organization.

Drawing our attention to how work processes are structured by computer technology, as life in the Bororo was structured by the plan of the village, structuralism deepens our understanding of what we are really up to when developing computer systems. Structuralism thus strengthens our belief in our own importance as change agents. But at the same time it offers us a solution to one of our fundamental puzzles, namely the contradiction between the systems we develop to ensure stability and the dynamic and unpredictable change processes of modern organizational life.

prac·tice (prăk′tĭs), *n.*, *v.* **-ticed, -ticing.** —*n.* **1.** habitual or customary performance. **2.** a habitual performance; a habit or custom. **3.** repeated performance or systematic exercise for the purpose of acquiring skill or proficiency: *practice makes perfect.* **4.** skill gained by experience or exercise. **5.** the action or process of performing or doing something (opposed to *theory* or *speculation*); performance; operation. **6.** the exercise of a profession or occupation, esp. law or medicine. **7.** the business of a professional man: *a lawyer with a large practice.* **8.** plotting, intriguing, or trickery. **9.** a plot or intrigue. **10.** a stratagem or maneuver. **11.** *Law.* the established method of conducting legal proceedings. [n. use of v., substituted for earlier *practic,* n.] —*v.t.* **12.** to carry out, perform, or do habitually or usually, or make a practice of. **13.** to follow, observe, or use habitually or in customary practice. **14.** to exercise or pursue as a profession, art, or occupation: *to practice law.* **15.** to perform or do repeatedly in order to acquire skill or proficiency. **16.** to exercise (a person, etc.) in something in order to give proficiency; train or drill. —*v.i.* **17.** to act habitually; do something habitually or as a practice. **18.** to pursue a profession, esp. law or medicine. **19.** to exercise oneself by performance tending to give proficiency: *to practice at shooting.* **20.** *Rare.* to plot or conspire. Also, **practise** for 12-20. [ME, t. OF: m.s. *pra(c)tiser,* ult. der. LL *practicus* PRACTICAL] —**prac′tic·er,** *n.* —Syn. **2.** See **custom. 3.** See **exercise.**

We tend to think of our work as engineering. Our conception of how technology brings on changes derives from our experience of how industrial technology has changed our natural environment. Computerizing an organization, we tacitly assume, is pretty much like building a highway or fertilizing a field: a matter of introducing the right cause in order to get the desired effect, a matter of good engineering. Structuralism wants us to replace this notion of engineering with that of tinkering, and it wants us to apply it both to ourselves and to our customers.

The systems developer has to be a tinkerer in relation to the material structures already in place in the user organization. Unless those structures are changed too, the introduction of a new computer system will scratch only the surface of the organizational culture. The systems developer has to accept the fact that users will apply tinkering to turn the technology they use into whatever they think it can best be used for. Material structures, such as computer systems, shape the life in an organization, according to structuralism, but human beings are not unwittingly accepting such external control. Systems developers consciously use computer systems to bring about change, and the customers of such systems will consider what kind of changes, if any, they want those systems to make possible.

If you want to play chess, you must accept the rule that a bishop can only move diagonally. A few rules like that together constitute the game of chess. There are other, socially mediated, rules regulating the way chess is played. Breaking them will often mean that you lose a game, but you will still be playing chess. This distinction between constitutive and regulative rules is more difficult to apply outside the world of games. But attempts to do so may very well be worthwhile.

Should politics be like a game with constitutive rules, or should it be more like a cocktail party with rules of a regulative nature only? Or, think again about the distinction between the stable and volatile properties of a system. One way of finding out which properties to treat as stable and which to treat as volatile would be to look for constitutive versus regulative rules in the organization.

Both critical theory and structuralism are *revolutionary* in the sense of not being afraid of attacking constitutive rules. Positivism and hermeneutics are both happy to work within the system, using knowledge of the system to perfect it and put it to good use. Positivism treats the constitution as natural laws on which to base predictions, explanations, and control. Hermeneutics treats the constitution as a framework for its dialogues. The framework may develop but cannot be overthrown. Critical theory advocates change – intervention aimed at changing the rules of the game – while structuralism wants us to see how difficult it is to achieve such change.

Designing computer artifacts, positivists will concentrate on functional qualities. Sometimes, they will turn this interest in the functional into an aesthetic theory, into functionalism, arguing that only what is functional is beautiful.

Hermeneutics are interested in art and aesthetic qualities, and prefer to turn their backs on technology, unless it is possible to give it aesthetic value. Their advice to designers of computer artifacts will be to rely on people rather than on technology.

Critical theorists are mainly interested in the political qualities of technology, in the way in which technology controls our lives and in how we can design it so as to make our lives rich and rewarding, rather than narrow and boring.

Structuralists stress the symbolic qualities of artifacts, wanting us to pay attention to their roles as symbol systems, communicating social organizations and ways of life. But structuralists also want us to see how the design of such artifacts is a process of tinkering, a game in which the leading role is played by the users of the artifact, even if the users seldom will be willing or able to change the rules of the game.

12

From Philosophy to Practice

Perspectives
Computer Metaphors
Professional Roles
Playing with Perspectives
Struggling with Quality

Dr. Pangloss was not a happy man. On his travels around the world with Candide, he met with terrible accidents and experienced unspeakable horrors. He lost the tip of his nose, an eye, and an ear from syphilis. He experienced the Lisbon earthquake, he was whipped and hanged.

All the evils, injustices, and miseries in the world could not make him change his philosophy: that everything that happens is for the best. Our world is well ordered and rational, a preestablished harmony based on the law of sufficient reason; it is the best of possible worlds. Dr. Pangloss knew that he had found the one true representation of the world. No alternative made any sense to him. Instead of changing his philosophy, in the end, he gives up philosophy altogether. With Candide, he goes "to cultivate our garden."

Dr. Pangloss is Voltaire's caricature of that great contributor to the mechanistic world view, the seventeenth-century philosopher Leibniz. We will not discuss how adequate it is as a caricature or criticism of that world view. Instead, we will try to see how we can avoid the fate of Dr. Pangloss, how we go about opening our minds to change.

Computers are very good at exposing mistakes in our thinking, especially those mistakes in which we think we do one thing and then do something else. When we test run our programs, we are typically surprised. We have no other technologies with a comparable, general capacity to expose mistakes in our thinking.

It is always difficult to evaluate ideas and plans without trying them out in practice. Sometimes we find out much later that our ideas or plans did not have the intended outcome. But often we deceive ourselves and stick to strategies and beliefs that are invariably contradicted by our experience. Or, we continue to preach one thing while practicing another. When we are not programming, the relation between thinking and action, between what we intend and what we achieve, between what we say and what we do, is obscure, and the dividing line between insight and deception is difficult to identify.

How do we know whether our design proposals, project plans, and strategies for improved productivity and quality will lead to satisfactory developments? How can we even tell if we, in our practice, act in accordance with the proposals, plans, and strategies that we say we follow? The only way we can do this is by paying close attention to how our plans and strategies relate to our practice. We must continually confront what we think with what we do. We must practice philosophy in action. But this is by no means easy.

If we manage, unlike Dr. Pangloss, to break out of our own philosophy enough to take an interest in alternative ways of looking at the world, and if we find that such alternatives are not just silly, then we will begin to compare the different perspectives, first, with each other and, later, with our practice. It is not easy to confront our philosophy with our practice. But we can do it by thus following a roundabout route, via a confrontation between our philosophy and other perspectives on the world.

In this concluding chapter we will deal with the fundamental challenge of our profession, the challenge to make the relation between philosophy and practice a dialectic process of confrontation and change, to turn the daily routine into a struggle for quality, to avoid becoming a Dr. Pangloss. In doing so, we will at the same time look back at some of the many perspectives on our profession introduced in this book.

Perspectives

Differences in perspective are a serious matter, creating misunderstanding and often unresolvable conflicts. Looking back on years and years of religious warfare in Europe, Leibniz blamed the conflicts on misunderstanding due to differences in perspective. Leibniz' attempt to construct an exact universal language, what later would become predicate calculus, was motivated by his belief in the value of everyone sharing the same perspective. Those supporting the mechanistic world view have a strong belief in the possibility of realizing Leibniz' dream, of arriving at one true representation of the world, of perfecting one scientific perspective on everything there is.

The romantics, on the other hand, regard the existence of differences in perspective as either an example of the richness of human culture or as an expression of more deep-rooted differences in interests. They consider Leibniz' idea of how to resolve human conflict as rather naive. They are not so much interested in reducing all the perspectives to one as to confront them, compare them, and combine them to enrich their view of the world, while still letting them go on developing each on their own.

The romantic concept of perspective has two aspects. A perspective is both what we see when we look at the world, from a certain point of view, *and* a way to look at the world, a possible point of view. With different perspectives, the world will look different. A perspective is a way of ordering the world, a simplification by exaggeration of the structure of the world, a highlighting of certain differences into fundamental distinctions. It is a conceptual framework of basic categories, a taxonomy enabling us to group and classify phenomena in the world, ordering them according to their relative value and importance. It is a few fundamental beliefs about what is and what is not, about what is important and what is not. As we use the term, a perspective can be anything from a simple point of view to a whole philosophy.

All professions engage in discussions and controversies involving different perspectives on their tasks, their clients, their products and services, and on the ethics and aesthetics of their profession. But as computer professionals, we have an additional reason to be interested in perspectives: Our tasks involve description and design as key activities. We apply and communicate perspectives when analyzing, designing, and discussing the systems we work with.

Perspectives can differ in at least three ways: standpoint, selection, and interpretation. Perspectives can differ in *standpoint.* If we look at a work situation, for example, we can view it from the outside, as it were, attending to general ways of doing things and on the relations between different tasks. But we can also view the situation from the inside, trying to experience what it is like to do the work with these colleagues. Positivism with its objectively detached view of reality and hermeneutics with its attempt to understand actions from the inside differ in choice of standpoint. One is a view from no-where, the other from now-here.

Perspectives involve the *selection* of certain properties and the neglect of others. Looking at a specific work process, we can focus on material flow or we can focus on information flow. We can selectively attend to how people use machines to manipulate materials and on the ways in which materials are transported from one work station to the next. We can also choose to observe how workers and supervisors communicate about work, disregarding for the moment the material flow.

When two persons disagree because of differences in standpoint or selection, the differences can often be overcome by providing more information. But when perspectives differ because of different *interpretations* of what is seen and experienced, providing more information may not help. When fundamental attitudes, values, assumptions, and interests differ, differences are difficult to reconcile.

This book introduces a number of perspectives from our intellectual history and applies them to the world of computers and systems development. Some are perspectives on social interaction and life in general. Others focus specifically on computers and systems development.

We began in part I with a discussion of two fundamental perspectives: the mechanistic and the romantic world views. In part II we presented three perspectives on systems development: construction, evolution, and intervention. And in part III we discussed quality from three different viewpoints: artifacts, culture, and power. In this last part we have introduced four theories of science – positivism, hermeneutics, critical theory, and structuralism – and we have used them to discuss the fundamental contradictions between computers and people and between systems and change.

All of these perspectives are intimately related to one another, and they have been introduced to support one another in deepening our understanding of the design of computer systems. To bring out the complexity of our practice, we have also introduced a number of more specific concepts: data, information, and knowledge; Platonic and Aristotelian concepts; bureaucratic and organic structures; tradition and transcendence; qualities and quality; technology, understanding, and emancipation; labor, language, and power; engineer, facilitator, emancipator, and tinkerer; and many others.

There are numerous parallels between all these different concepts and perspectives, all going back eventually to the fundamental distinction between the mechanistic and romantic world views, passing on the way perspectives on systems development and theories of science. From language, to take only one example, we come to understanding, to the facilitator, to organic organizations, to Platonic concepts, to evolution, to soft systems, to hermeneutics, and to the romantic world view. Doing this for all the concepts and perspectives can be a good exercise to see how much you remember of the discussions.

It makes sense to think about the relations between all the different distinctions or taxonomies introduced. For instance, we have introduced the distinctions between construction, evolution, and intervention, and based on Habermas' ideas we have later introduced the distinctions between the engineer, the facilitator, and the emancipator. In many respects these two taxonomies are similar. Engineers are primarily interested in technology, and they understand their task and approach as the construction of the optimal technical solution to be delivered to a client in response to given requirements. Facilitators are interested in understanding. They engage in a dialogue with clients and take an experimental, evolutionary approach to develop satisfactory computer solutions. Emancipators are critical and concerned with power. They involve themselves directly in the problematic situation of the client, intervening to change the existing power structures and to create a more equal distribution of resources and opportunities. But if these distinctions are so similar why don't we just select *one* taxonomy, define its categories clearly and thus simplify our discussions?

Bringing together perspectives from different sources, we have to remember that each of the taxonomies are developed within a specific context and for a specific purpose. The distinctions between construction, evolution, and intervention are developed in response to a key question: What is development, in particular the development of computer systems? Each of the three concepts provides a different view of what the task is and of how to approach it professionally, and they all view the agent performing the task as an individual person or a small team.

The distinctions between engineer, facilitator, and emancipator are based on Habermas'theory of the different media of social organization and his related ideas about different knowledge interests. We have adopted Habermas'ideas to discuss the issue of interest and power in relation to systems development and computers. The two taxonomies are really very different and difficult to merge. We have tried to provide enough of the original contexts to make this clear, but we are aware of the difficulty of the task.

Our ambition has been to provide all these different taxonomies in order to enrich and make more complex our understanding of systems development, not to show that there is really only one simple dichotomy being varied over and over without much of anything new being added. The purpose of this book is to help you compare different taxonomies and to see how they are related. The goal is not to memorize general and superficial schemas but to be able to get a handle on the complex nature of our profession. We want you to get on with a creative discussion of different perspectives in order to challenge and develop your own *personal* view of our profession, hopefully by going on to find other perspectives, correcting some of our simplifications, simplifying some of our complications.

General perspectives like the mechanistic and the romantic world view tend to degenerate into clichés. They divide the world into separate, isolated areas: Computer technology is an expression of the mechanistic heritage in our culture and the romantic challenge represents the human dimension of computing. Here, we have the mechanistic world with computers and technology. And over there, we have the romantic world with people and organizations. We don't want to encourage such stereotypical use of these perspectives.

If computers are not simple expressions of our mechanistic heritage, what then are the romantic aspects of computers? Can we use a romantic perspective to enrich our understanding of computers? If the human dimension of computing is not simply an expression of the romantic challenge, what then are the mechanistic elements of our professional practice? Can we use a mechanistic perspective on, say, a software project team and turn it into a powerful machine without just falling prey to the silly simplifications ridiculed by Gerald Weinberg? In the following sections we shall try to answer these questions by discussing a few concrete and fruitful perspectives on the two basic elements of our profession: computers and people.

Computer Metaphors

The first computers were built to mechanize the work of the human computer. They were constructed to repeat over and over again a variable but simple pattern of instructions chosen from a small set of arithmetic operations. These first computers were really *computing machines*. But already in those early days of computing, people disagreed on the potentials of this new technology.

Howard Aiken could not imagine how computers could become useful artifacts for administrative work, while John McCarthy believed that these machines could be designed to perform any kind of intelligence task. We know now that Aiken's view of the computer as a computing machine was much too unimaginative. We have a long way to go before realizing, if it will be at all possible, the vision formulated by McCarthy. But computers are no longer, in any reasonable sense of the term, just computing machines.

Many computer systems can be viewed as *media*. A computer-based production control system is a medium for communication between planners and supervisors, between supervisors and workers, and between individual workers. Viewed as a medium, a production control system provides the actors with opportunities for communicating and negotiating plans, problems, and breakdowns.

Electronic mail is one of the most obvious and powerful applications of computers as media. Here, the computer is not used as an advanced calculator to help us compute various functions. Instead it is used as something between a telephone and a traditional mail system.

The use of electronic mail resembles the use of the telephone. There is a fairly quick exchange of messages and the style is typically informal and spontaneous. But there are also differences: One medium offers written communication, the other oral; one medium enables the receiver to respond whenever convenient, the other is more like an interrupt, requiring an immediate response.

Correspondingly, we can look at similarities and differences between electronic mail and traditional mail. From such comparisons, we see that computers have qualities that we know and appreciate from other kinds of media. These qualities make them extremely useful as a means of communication.

In other cases it is more fruitful to think of the computer as a *tool*. Using the tool metaphor, we stress the relation between the computer and the purpose for which it is used. We introduce the norm that good computer systems should disappear into the background. When a person is using a computer to write a text, she should be concerned not with the text-processing system but with the text itself. When programmers are using a programming environment, they should not be concerned with the concepts and facilities it offers but with the problem at hand and the program to be produced.

Using a tool metaphor, we stress the primary task of the users, and we try to design computer systems that assist them in performing these tasks in a professional and effective way. Looking at computers as tools forces us to go beyond a technical, computer-centred notion of quality. We are encouraged to emphasize quality of use.

The tool metaphor raises many important questions. In designing a specific computer system, what is the trade-off between easy learning and effective use? How can we avoid breakdowns and assist the user in concentrating on the primary task? How can we design the system to react sensibly when breakdowns occur?

Other relevant metaphors can be suggested. The computer is also an *automaton*. We use computers to automate intellectual processes. We integrate computers into other technical systems and we use them to develop robots and programmable production technologies. What are the important characteristics of the computer as an automaton? What is a good automaton? How can we define quality as related to automation?

Some computer metaphors have mechanistic origins whereas others are more romantic in nature. If we view computers as computing machines or automatons, we are stressing mechanistic ideals. We view computer systems as artifacts abstracted from the context in which they are used. Computing machines receive inputs, perform functions, and deliver outputs. Automatons perform functions that substitute for human information-processing activities.

If we instead view computers as tools or media, we are stressing romantic ideals. Both metaphors emphasize the relation between the artifact and its social application. Viewed as a tool, a computer system is subordinate to the professional activity for which it is designed; viewed as a medium, a computer system is seen as a facility and framework for human communication and interaction.

Computers are not just computers. If we maintain a simple and one-dimensional conception of computers, we will also have a simple and rigid understanding of the problems and possible solutions of our professional practice. To advise users and clients on when and how to use computer technology, we need a rich understanding of the potentials and limits of computer technology. We must be able so see the computer from many perspectives, and we must be aware of how these perspectives result in different, and often conflicting, demands on quality. A good automaton is not a good tool, and vice versa.

Practicing philosophy in this way, we move from the general and abstract distinction between mechanistic and romantic ideals to more specific conceptions of computer systems. We try different perspectives and we move closer to our own practice. Still, with such intellectual exercises we remain detached from the concrete complexity and excitement of our professional practice.

Electronic mail is a good example of the use of computers as media. But a discussion such as the above is still a poor and rigid rendering of our everyday contact with electronic mail systems, which might instead sound something like this:

"I like them. And I hate them. I use them to communicate with my closest colleagues and friends. We exchange news and gossip, we negotiate and make plans, and we remind each other about promises and arrangements. This is all very effective and pleasant. And it works whether your colleague is sitting next door or across the Atlantic. But then I receive too many messages that I don't care about and that require me to do something I would rather forget. I have also started to wonder about the negative atmosphere of the mail communication at my department. As chairman, I am involved in many controversies and discussions on the mail system. There is a lot of negative criticism. A lot of complaints. Rarely do I find any positive remarks. People blame each other for what they do wrong or didn't do rather than praising the things they did well. Why is this so? Is email an unhappy mixture between the more formal, traditional letter and face to face confrontation and discussion. Email invites you to practice an informal style, but it does not require you to look into the other person's eyes. Is it too easy to be negative in this medium? Well, I know for sure, that I would like the system to automatically delete more than half of the messages I receive!"

What do you think? How would you like people to use email? Would you prefer a different kind of informal communication system? Is change still possible?

Professional Roles

Computer professionals must master a wide spectrum of methods and technologies. They must have the energy and skill to quickly learn and evaluate new technologies and to modify and extend their repertoire for action. They must be able to cope with unstructured situations, understanding and appreciating the unique and specific characteristics of the problematic situations involved in their daily work. Often they have to work under more or less formal, contractual arrangements with limited resources. But at the same time they must be able to interact with imaginative and demanding users and managers in quite turbulent environments.

None of us are able to cope with all these different demands in systems development. People are at least as different as com-

"He claims to be a specialist, but I think he just has a
one-track mind."

puter systems. Some of us are more analytic, some more creative. Some are successful as project managers, others are competent programmers. Systems development requires effective cooperation and the ability to benefit from and develop all the different experiences and qualifications of the involved actors. But how can we achieve this? How can we understand, appreciate and effectively manage the differences between individual systems developers?

Reflecting on these questions, we could rely on our understanding of activities and competencies related to systems development. Any project requires management, analysis, design, programming, and testing. It also requires cooperation, communication, negotiation, and conflict resolution; and it requires a variation of skills related to the technologies involved – for example, programming languages, CASE tools, operating systems, and data base management systems. These competencies are, of course, highly relevant, but the distinctions are rather specific and their relations complex. In addition, there is the danger that we learn very little by relying on what we already take for granted within our profession.

Instead of drawing on our practical experience of systems development in discussing the role of individual differences, we have chosen a general theory of the different types of professionals that easily applies to systems development. According to this theory, there are four roles that are necessary and sufficient to develop effective professional teams: the producer, the administrator, the entrepreneur, and the integrator. Each role represents a perspective on computer professionals and the qualifications of each individual systems developer can be understood as a specific mix of these roles. A computer professional may be well qualified for some of these roles. For other roles, the same individual may be quite hopeless.

The *producer* is effective in producing results and services. Producers master the key disciplines of analyzing, designing, and implementing computer systems; and they are competent users of the methods and tools involved. Producers have strong motivation to make things work. Systems development projects need producers to be able to deliver the systems and services required. But motivation and technical competence are far from sufficient to ensure success.

The *administrator* is effective in controlling the development process so that the agreed-upon results are produced. Administrators make detailed plans, they coordinate and control, and they set up rules and enforce procedures. Administrators make sure that the process develops according to plan. Systems development projects need administrators to take care of the complexity of the·process and to ensure discipline and progress. Project planning, status reports, and configuration control are well taken care of in the hands of an able administrator.

But successful systems development requires more than effective production and administration. Systems development projects face uncertainty and the challenge to change and transcend traditions. The *entrepreneur* is effective in generating and initiating ideas concerning the organization of the process or the design of the system. Entrepreneurs are self-starters. They are creative and prepared to take risks in testing new opportunities. Entrepreneurs participate actively in establishing and evaluating goals and strategies. Systems development projects need entrepreneurs to change approaches and suggest novel solutions.

Even if the productive, administrative, and creative abilities are there, projects may still fail. A key condition for successful systems development is that the project group is able to operate as a team. The *integrator* is effective in creating a well-functioning project group out of individual differences. Integrators transform individual risks to group risks, they establish harmonic relations between individual interests and project goals, and they make individual entrepreneurs part of a cooperative effort. Integrators focus on group dynamics and on personal and professional development. Systems development projects need integrators to make users and systems developers communicate and cooperate and to facilitate effective cooperation among systems developers, taking into account their different preferences and working habits.

Using this simple theory, take a moment to reflect upon your personal qualifications and preferences. How much of a producer (P), an administrator (A), an entrepreneur (E), and an integrator (I) is there in you? You can challenge your self-knowledge and self-deceptions by considering some extreme cases.

Lone rangers have a strong P and a weak A, E, and I. They produce results, they are always busy, and they focus their attention on the things they are working on right now. Lone rangers judge themselves by how hard they work. When they make decisions, they act first and think afterward. Systems development projects that are dominated by loners are likely to deliver technically sound computer systems on schedule. But the production process is chaotic and everyone works hard. There is a considerable risk that the use quality of the resulting system is low. The system will need to be modified or changed to suit the actual needs of the users.

Bureaucrats have a strong A and a weak P, E, and I. They like to keep order, and they like to engage in controlling progress and results. They work systematically, and as a consequence they are slow but careful. Bureaucrats are conservative and anxious to abide by decisions already made. Systems development projects that are dominated by bureaucrats have a smooth and well-organized process. The atmosphere is quiet. The bureaucrats make sure that the delivered system is in accordance with formal requirements, but it is more than likely that the resulting system is at most a partial solution.

Arsonists have a strong E and a weak P, A, and I. Arsonists are innovators and entrepreneurs. They are always concerned with new things, always enthusiastic, stimulating, and creative. They try to engage others in their ideas. Arsonists are eager to participate in making new decisions, but their commitment is low, and they tend to remain in the world of ideas. When arsonists dominate, projects will typically start well and with lots of enthusiasm. But as time passes, the project will fall apart and no system will be delivered. Instead new projects will be initiated.

Super followers have a strong I and a weak P, A, and E. They are concerned with making people agree. They suggest compromises and integrate ideas suggested by others. Super followers are oriented toward people, and they are understanding and focus on mutual agreement. Super followers are willing to make decisions only when everyone agrees. A concentration of super followers will ensure a good atmosphere in the project. The users are satisfied because their demands are seriously discussed and never rejected. But there is little chance that anything new comes up, and it is more than likely that the project will not meet its deadlines.

The four roles – producer, administrator, entrepreneur and integrator – correspond to four different views on systems development: systems development as production of computer systems; systems development as planned and organized work; systems development as design and invention; and systems development as a cooperative and social process. The claim is that each of us is more or less willing and able to take each of these views in our profession. But for a project to be successful, all four views have to be represented.

Reviewing the four professional roles from a mechanistic and a romantic point of view, we make the same kind of observation we made when reviewing computer metaphors. Focusing on people, emphasizing the human dimensions of systems development, does not restrict us to romantic ideals. Instead, a different kind of pattern seems to emerge, emphasizing again the relations between mechanistic and romantic ideals.

The very idea of roles is mechanistic. When we use the four categories to select the members of a project team, we think of the team as a machine and of its members as the functional units of the machine. But on a more specific level, two of the roles are examples of mechanistic ideals, whereas the other two exemplify romantic ideals. And from another point of view, two of the roles focus on product issues, whereas the other two focus on process issues. The lone ranger and the bureaucrat are mechanists, the former concerned with the construction of computer systems and the latter with organizing and managing the development process. The arsonist and the super follower are, in contrast, more of a romantic nature, the arsonist busy designing new systems and the super follower deeply involved in the social aspects of the development process.

Are you a lone ranger? Or are you more of an arsonist? Are you happy with your professional style as you see it? Do you know how other people view you? What possibilities do you see for further professional development? Is your project doing well, or is there a need to complement the present mode of operation with other qualities? Is your project group a well-functioning team? When we try to answer these questions, we begin something like this:

"I have always found it difficult to characterize and evaluate myself. My immediate impression is that my colleagues like me and that they appreciate my contributions. Well, I'm definitely a strong producer and entrepreneur. I usually work hard on producing results, and I'm also inventive and creative in forming new proposals and suggesting possible solutions. Am I a good administrator? Well, at least I worry about the project group. I talk to my colleagues and it was Eleanor and me who organized the last party for our group. I'm socially concerned and I care. There must be something of an integrator in me. But what about administration? Am I too lazy to do that? I know for sure that I prefer John to write up all the project plans and minutes of our meetings. By the way, I don't care much about these minutes. Couldn't we do without them, and get on with the project? I am definitely a systematic and hard-working producer. Why don't I then really care about administration? Maybe I have a much too positive view of myself? I wonder how Eleanor sees me."

Playing with Perspectives

We use perspectives all the time. Our language and thinking are suffused with metaphors expressing particular perspectives. Just think of the way we describe ordinary objects in our immediate environment: chairs have legs and arms, mountains have foots, clocks have faces, books have backs, we move the cursor with a mouse, and computer systems offer windows, documents, and folders. Or think of how we use geometry in our thinking and speaking about emotions and careers: I am feeling really low today; drink this and you will get high; he is down and out; she is coming up in the world.

But most of these metaphors and perspectives are dead. We no longer think of them as metaphors and we see nothing strange in speaking or thinking this way. These metaphors have become integral parts of our daily practice that we take for granted. In contrast, the idea of this book is to explicitly play with perspectives and metaphors to stimulate learning.

Imagine that you are engaged in designing a new project for developing a management information system for a university. While being so engaged, you may take time off your immediate concern to ask yourself what university education is all about. Using metaphorical thinking, playing with perspectives, you will come up with a number of answers just by comparing a university to other institutions.

A university is a hospital. You go to a hospital when you are handicapped by illness. The doctors treat you, and you leave the hospital a more capable person. You go to university because you are handicapped by ignorance. The professors treat you, and you leave the university a more capable person. A university is a factory. In a factory raw material is processed in a production line, leaving the factory as a finished, standardized product. Students are the raw material for universities. Passing through a curriculum, they are worked upon by professors to leave the university as standardized products. A university is a prison.

Inmates in a prison are subjected to discipline, taught to behave themselves. Students are taught by professors to sit down and listen, to behave in accordance with the norms of our society. A university is a sports event. Students compete. Some do better, some worse. They are expected not to cheat. Professors judge their performance, ranking them, handing out prizes.

Reflecting upon these different perspectives, you should ask yourself certain questions: What is the role of information in managing university education? What is the present and potential use of formalizations and computing? What kinds of strategies and methods should be used in developing a new management information system?

Playing with perspectives in concrete terms like these, or reflecting in more general terms upon the mechanistic heritage and romantic challenge of our profession, might seem just like a game to you. Systems development is not *play*, you might say, but hard *work*, thus exemplifying a fundamental pair of opposing perspectives in our culture. Is this for real, or is it just play? Are we doing serious work here, or are we just playing? Are we closing in on the truth here, or are we just playing with words?

But that opposition is spurious. Children need to play in order to learn, effectively, about the world. To children, play is serious work. One way of seeing why it is so difficult to build bridges between a mechanistic and a romantic world view is to see how they differ with respect to this opposition between work and play. The difference between science, objectivity, truth, and absolute standards of quality on the one hand and art, subjectivity, perspectives, and quality as transcendence on the other is conceived by the mechanist as the difference between serious work and just play.

If you look back on the way we have talked about systems development and computers, you will find that we just go on and on about work. We bet you did not think twice about this, in spite of the fact that it is not unlikely that the most economically significant use of computers today is for play. We tend to think of modern society in terms of work, work, work, in spite of the fact that the fastest growing industry in the twentieth century has been the play industry, the production of entertainment of all sorts. Why doesn't Habermas include play among his media or means for social organization? How about war, you may wonder. Isn't that where the real computer money is? Well, how about it? Is it work, or is it play? Habermas would of course view play as communication, and war as power.

To philosophize in systems development is to search for knowledge about the nature, meaning, and practice of our profession. It is to participate in increasing our knowledge of the inherent challenges of this specific kind of human activity. As philosophically minded systems developers, we reflect on our practice and in doing so we develop a personal view of our profession. This book is based on the romantic idea that systems developers can improve their practice by deliberately playing with perspectives. But the play is not a frivolous pastime. We do not play without rules, and the objective of the play is to develop a philosophy, integrated with our practice, that we can fully commit ourselves to.

Struggling with Quality

Frederick Brooks gives us a thought-provoking historical analysis of software engineering. There is no simple solution to our problems, he says: "Of all the monsters that fill the nightmares of our folklore, none terrify more than werewolves, because they transform unexpectedly from the familiar into horrors. For these, one seeks bullets of silver that can magically lay them to rest. The familiar software project, at least as seen by the nontechnical manager, has something of this character; it is usually innocent and straightforward, but is capable of becoming a monster of missed schedules, blown budgets, and flawed products. So we hear desperate cries for a silver bullet – something to make software costs drop as rapidly as computer hardware costs do. But, as we look to the horizon of a decade hence, we see no silver bullet."

Using Aristotle, Brooks distinguishes between the essential and the accidental difficulties involved in software engineering. Some of the difficulties are inherent in the very nature of software, and we shall have to live with them as long as we continue to work with software. Other difficulties are accidents of the historical development of our practice, and those difficulties can be removed, if only we can identify them and understand their origin.

The inherent properties of software are, says Brooks, complexity, conformity, changeability, and invisibility. Past technological breakthroughs, like high-level languages, time sharing, and unified programming environments, attack the accidental difficulties related to the tasks of *representing* the software and testing the fidelity of the representation. None of them attack the essential challenge: the specification, design, and testing of the *conceptual constructs* underlying the various representations. Evaluating contemporary technological developments, such as object-oriented languages, expert systems, CASE tools, and graphical programming, Brooks reaches a similar conclusion.

Brooks' analysis addresses the contradiction between technology and people, and he concludes that our technological innovations do not attack the essential difficulties of developing software. His advice to software managers responsible for future investments and strategies for change is simple: Buy and modify standard software rather than developing your own; when you have to develop software yourself, experiment with the fundamental conceptual constructs through requirements refinement and prototyping; do not build software, but grow it by using an incremental strategy; and, finally, hire skillful designers and provide incentives for them to develop their skills.

Brooks' argument can be questioned, but the conclusion of his historical analysis is consonant with what we have learned from philosophizing on computers and systems development. The design of computer systems demands a mechanistic attitude.

Without technical skill and technical fascination, without persistent attempts to obtain at least a transient stability, order, and conceptual simplicity, there is little hope that we will ever achieve anything, not to say deliver a functioning computer system.

But, at its heart, systems development requires a romantic approach. The challenge is to cope with people and with the dynamics, dilemmas, and power games involved. Without a romantic attitude, computer professionals degenerate into naive and narrowminded experts defending their position by insisting on rational and technical ideals detached from the realities of organizational life.

This is all very well, you might say. But what happens to these general ideals when they are subjected to the reality of specific organizational, economical, and technical traditions? What if the conventional wisdom at the workplace is quite different? What if the actual possibilities for experiments and change are extremely limited? The easy way out is, of course, to find a different job. And this does in fact seem to be the thing to do in our profession. But in most cases, such a transformation will only give you an opportunity to engage once again in the same old game. Rarely will such a move imply a fundamental change.

The very idea of systems development is to change organizations. But to really change a social organization is difficult, structuralism will tell us. And to change an organization in the direction intended, to improve its quality, is even harder. It is as difficult as changing a person, as changing yourself. But unless we want to end up as Dr. Pangloss, cultivating our gardens rather than our competence as systems developers, we must learn to change – both our practice and our philosophy. And the only way this can be rationally done is by forcing our philosophy to confront our practice, and our practice to confront our philosophy.

Notes for Instructors

The obvious way to use this book is as supplementary reading in one of the traditional courses on computing, software engineering, or information systems. It can then be assigned for self-study, but you may want to bring up some of the concepts and themes of the book in lectures as a framework for discussing and evaluating the main theories and methods taught in the course. We have worked hard to make the book clear and easy to read in order to facilitate this simple way of integrating philosophy into your curriculum.

A more ambitious use of the book would be to include it as compulsory reading in a course on computers and society, theoretical foundations of computing, software engineering, or systems development methods. The four parts of the book constitute a framework for a series of four lectures, with exercises to encourage student discussions. In this case, a more ambitious effort should be made to relate the themes of the book to the rest of the material within the course.

The most ambitious way to use the book is to have a specially designed course on philosophy for computer professionals. We are well aware that this, in many cases, is a utopian idea, far from the realities of today's curricula. But in our experience, this effort is truly worthwhile, and students will eagerly engage in discussions about their education and future practice. In this case, each of the chapters provide a framework for one lecture, again with exercises to encourage and organize student discussions. In teaching this kind of course in Scandinavia, we have found it fruitful to complement the book with practical books on argumentation and style – for example, Weston (1987) and Strunk & White (1979). In our own courses students have written essays expressing their own views on some of the issues raised in the course.

While writing this book, we used three questions as fundamental criteria of quality for each chapter: Would a teacher of software engineering or information systems like to give a lecture based on this chapter? Could you base an examination on this chapter? Are the relations between this chapter and the others clear enough? When teaching a seminar, the exercises can be used as a basis for discussion, and students can be challenged to use them for further reflection. When giving a series of lectures, the titles of the subsections can be used as a checklist, and the exercises can be used both for student homework and for a final examination.

Any interdisciplinary effort runs the risk of being reduced to the superficial common denominator of the disciplines involved. In writing this book, we have tried to avoid this by relying on our professional backgrounds in computer science and philosophy. But when teaching the book, you will probably not have a colleague with a complementary background at your side. How then can you teach a course on the book without having to study background material with which you are not acquainted? Is it sound teaching practice, for instance, to teach a course on this book if you have never heard of hermeneutics or structuralism before? Our answer is yes.

Teaching a course based on this book should present no serious problem for teachers trained in computer science or information systems. The fact that the discussions in the book are always related to programming, analysis and design of computer systems, project management, quality assurance, and other areas within the teacher's professional competence, provides the necessary firm ground to begin from. If students want to go beyond the book and learn more about hermeneutics and structuralism, they can pursue self-study, or a guest lecturer can be invited to expand on the issue.

Conversely, if the book is used by teachers with no training in computer science or information systems, it is possible to begin from the other end, concentrating on the discussions of systems thinking, the theories of technology and social change, the theories of science, and so on. The student who wants to know more about structured techniques and object-oriented thinking will in this case have to go elsewhere.

In both cases it is important for instructors to make clear to the students where their competence lies, what they know and do not know. Indeed, teachers who are not afraid to thus admit to a class their ignorance in a field in which they are not educated are given an opportunity to have the unusual experience of joining the class in learning in the classroom. And the students will, of course, have an equally interesting experience of working more closely with the teacher. We are not saying that it is easy to participate in such a learning experience, but the results are rewarding. After all, when education is easy, no one is really learning anything.

There is very little time explicitly allotted to philosophical reflection in the education of computer professionals. One reason is perhaps that much of that education takes place either at schools of engineering or at business schools where philosophical reflection is viewed by many as a luxury. It is our conviction that even the smallest amount of such reflections will increase the quality of education substantially and help us educate more competent practitioners.

Philosophical reflection on the foundations of our disciplines and professional practices is a truly interdisciplinary effort where we have to take intellectual risks to gain practical relevance. As teachers, we must, of course, make sure that our students do not substitute a lot of philosophical hype for detailed and solid academic knowledge. The way to do this is to ground all philosophical discussions in concrete examples and well-understood theories. On the solid platform of our academic competence or practical experience, we must dare to raise philosophical questions without feeling any obligation to manage the jargon or give a lecture on Heidegger.

Philosophy has its theories, concepts, and methods. But if our interest lies in philosophizing rather than in philosophy, we need surprisingly little of the technical jargon philosophers use to express their theories, concepts, and methods. We can use those methods, learn from those theories, and make the relevant distinctions without having to introduce the jargon itself. It makes good sense to avoid too much philosophical jargon. Our intention is not to become philosophers but rather to challenge ourselves to become more professional.

A philosophical discussion of systems development is an invitation to reflect upon our practice. It is an invitation to a rational conversation with arguments, counterarguments, definitions, analyses, and conclusions; an invitation to discuss presuppositions and assumptions that we in our day-to-day work take for granted.

A philosophical inquiry typically starts by asking "What is this?" with a very special tone of thoughtfulness. In order to answer a question like "What is systems development?" there are a number of philosophical methods available. Let us give five examples.

In *linguistic analysis* we look at the way we describe systems development, at the words and concepts we use rather than the activity itself. A linguistic analysis naturally turns into a closer look at the concepts system and development, and an analysis of those concepts brings up other concepts such as structure, process, design, analysis, management, project, and product. In practice a linguistic analysis is a useful approach to understand and evaluate project models, methods, standards, requirements documents, and design proposals.

A *phenomenological description* focuses on systems development as it is performed and experienced in concrete practice.

The goal is to give a description of the practice that is true to the practitioner's experience. A good phenomenological description is an invaluable basis for theoretical and normative attempts to go beyond current practice. Phenomenology is, of course, important in everyday work, even if it is practiced without such a fancy name. Changing the operation of a software house requires an appreciation of present practices, and effective project management requires insight into and a feel for the working practices as well as the current state of activities within the project.

In *contextual analysis* we look at the larger framework of which systems development is a part. We look at the economic, technological, and organizational conditions for doing systems development. We inquire into the relations between systems development on the one hand, and decision making, organizational learning, power games, and organizational change on the other. This kind of analysis is needed to understand the conditions for systems development activities.

When we use *analogies*, we analyze systems development by comparing it to other human activities. Systems development can, for instance, be viewed as technical construction, as human learning and problem solving, as communication and negotiation, or simply as a political process. Using analysis by analogy, we can transfer experiences and methods from other professional fields, thereby enriching our own repertoire. In specific projects, analysis by analogy is useful to develop reliable and convincing estimates and, also, to support us in reusing previous design ideas in new settings.

In a *historical analysis* we attempt to understand systems development by studying its history. We try to understand how it evolved into its present form and to see clearer what it is today, in all its complexity, by seeing how it has grown out of more simple, less sophisticated practices. We tend to forget our history, both as a profession and within the organizations where we work. Historical analysis helps us understand our successes and admit our failures. In considering whether to invest in new tools and methods or in designing new organizational settings for our projects, we might do well to first evaluate previous investments and change processes.

When we begin to philosophize, there is always the danger that we will end up in an abstract discussion with little or no relevance to the business at hand. If the aim is to become more professional in our practice, questions of framework and choice of perspective will have to be related and subordinated to concrete problems of practice.

One way of ensuring that our philosophical reflections do not stray too far from our subject matter is to postpone all worries about perspectives and to begin with concrete examples, real worries, and to attack them with a more direct approach. Without explicitly having to think about methods or perspectives, we can at any time stop and ask ourselves certain questions: How do I speak about my tasks, my colleagues, my clients, my job? Do I use jargon to shut out clients, to establish a professional image, or as a genuine means for better communication? How do I spend my working day? What kind of building, office, do I work in? What does my material environment reflect about my work? Who is my employer, my client? What are the most important concepts I use? What are my sources of information? What are my ideals, my dreams? What would I really like to do? Why do I not do it?

The exercises in the book are invitations to work with this kind of very concrete questions. You can approach the exercises by applying one or more of the philosophical methods mentioned above. But a more direct approach is to start from the few questions in the exercises and then infer from them more specific and personal questions.

Exercises

Chapter 1

1. Give examples of rules taught in your programming courses. Do you follow those rules when you program? Are there other rules you follow? Is programming a rational, rule-governed activity?

2. What is the role of mechanistic ideals in your education? Give examples and discuss their relevance.

3. Think of a game you like to play. Discuss the formalization involved and the mechanisms used to enforce the rules. How do you change the rules? Describe the use of a computer system as if it were a game.

4. How is the use of computers organized at your workplace? Is the balance between bureaucratic and organic elements appropriate?

5. How do you organize your own use of information? What are the mechanisms you use to store, retrieve and process information? When and why did you last change these mechanisms?

6. How do we use formalizations in organizations? Why and when are organizations formalized? What are the more general advantages and disadvantages of formalization?

Chapter 2

1. How do you collect and use information to improve your professional competence? Is your performance effective? How can you improve it?

2. Consider the following game. Each player is given a piece of paper with five columns, representing five categories: girl, tool, capital, famous person, and furniture. A letter is randomly chosen, and each player must write down a word beginning with that letter, for each category. As soon as someone has completed that task, all players must stop.

   Points are now given by considering and evaluating the words submitted by each player. If the group agrees that a word belongs to the category, the player gets one point. If this word is unique – i.e., no other player has chosen this specific word – the player receives one extra point. When all words have been evaluated, the players proceed to a new round.

   After playing the game, think about what you learned about the linguistic habits and use of concepts in the group. What did you learn about the categories?

3. Consider a method for database design or a method for analysis and design of computer systems. Discuss how this method deals with the formalization of information to data, the interpretation of data, and the transformation of Platonic concepts to Aristotelian ones.

4. Are the concepts of algorithm, database, and computer inherently mechanistic? To what extent are we as software developers and users of computers forced to adopt a mechanistic world view? Do we have romantic options?

5. Read one of the following papers: Boehm (1988), Floyd (1987), Humphrey (1988), Parnas & Clements (1985). Characterize and evaluate its mechanistic and romantic aspects.

Chapter 3

1. Describe the use of computers at your own workplace with a hard systems, a soft systems, and a dialectic systems approach.

What are the major features and challenges identified by each of the three approaches?

2. Compare and contrast the strengths and weaknesses of the three approaches to systems thinking.

3. Are the contradictions that play such an important role in the dialectic systems approach already attended to, as different perspectives, in the soft systems approach?

4. Think of your private life or your professional situation. What are the three most important contradictions? Which of these do you regard as temporary, and why? How do you deal with the others?

Chapter 4

1. Evaluate your last programming experience. Identify the elements of systems construction and propose improvements based on this approach. Give examples of events and conditions that made it difficult or impossible to follow a systems construction approach.

2. Design and implement a computer program to solve the 8-Queens problem. Record the way you work, or if you do it in a group, have someone in the group keep a record. Evaluate how you did it and compare your experiences with the description provided in Wirth (1971).

3. Discuss the distinction between top-down and bottom-up, and relate it to part-whole, abstract-concrete, specific-general, and analysis-design.

4. Discuss the technical, economical, organizational, and personal conditions required to practice systems construction.

5. Take any method for the analysis and design of computer systems. Does it begin with a well-defined data processing problem?

6. Go to the library and get some books on problem solving. Look for useful heuristics for general problem solving and compare the use of such heuristics with the way problems are solved in the systems construction approach.

Chapter 5

1. Give examples of significant errors from one of your projects. How did they affect the success of the project? Could they have been handled differently? Could they have been avoided?
2. What are the strengths and weaknesses of prototyping?
3. Using the theories of systems construction and evolution, describe and evaluate the way you planned and experienced your last vacation. Compare this with the way you planned and experienced your last work project.
4. Compare and contrast construction and evolution, considering economy, project planning, management oversight, cooperation with clients, change control, qualifications, and team building.

Chapter 6

1. Give a couple of examples of breakdowns in your use of a specific system and describe in detail how you reacted. What can we learn from these examples about the quality of the system and about your qualities as a user?
2. Consider important information-processing aspects of your own life, and identify structured domains for which you could use a computer. Would you really want to use a computer for these purposes?
3. For a designer, the artifact is naturally present-at-hand. For the user, the artifact should ideally be ready-at-hand. What problems does this give the designer? What methods could be used to handle these problems?
4. Identify major changes in patterns of behavior in your project group (family). Were these changes the result of explicit interventions from group (family) members? Are you satisfied with your own role in these processes?
5. Think of some interviews with a consultant, be it a colleague, a friend or a professional consultant. What did you like and what did you miss? What does it take to be a professional consultant?

Chapter 7

1. Describe the symbolic, aesthetic, and major functional qualities of your favorite computer system.
2. Define the concept of quality, using first an Aristotelian approach and then a Platonic approach. Discuss the complications involved in defining quality.
3. Evaluate a user manual. Does the manual include evaluations of the system? Should it?
4. Examine a systems development method. What does it have to say about the relation between the quality and the qualities of the system? Does the method include evaluation as opposed to description and specification?
5. Describe and evaluate the dominating view on quality and the mechanisms used to evaluate performance in your workplace?

Chapter 8

1. Discuss the strategy for improving the quality performance of the software house in chapter 8.
2. Is self-deception a prominent characteristic of the quality culture in your organization?
3. Identify and evaluate the most characteristic systems of signs (architecture, documents, methods, rituals) in your own organizational culture. What kind of values, ideologies, and assumptions do these sign systems communicate?
4. Imagine applying the strategy of the Salesian missionaries to your own organization. Identify and describe a single structural change that would create an opportunity for changing the ideology, of your organization.
5. Examine a book on quality assurance. Evaluate its notion of quality, its approach to quality assurance, and its view on strategies for changing and improving the quality assurance activities in an organization.

Chapter 9

1. Describe and evaluate yourself as a user of computer systems. What are your preferences, habits, and routines? Do you make effective use of the facilities offered by the systems?

2. Assume that you are about to develop a large administrative system, where the typical user will be someone like yourself. What consequences will this have for your approach?

3. You are offered a new and attractive job, which will mean a 50 percent increase in income. The professional environment is challenging, and the working conditions are superb. The company offering you this job is involved in arms development, but you are to work in a project without any relation to this. Will you take the job?

4. Evaluate the way computer usage is organized in your own department. Are the physical and psychological working conditions good? What kinds of services are provided to support users? Identify possible and necessary improvements.

5. What does it take to be a successful engineer, facilitator, and emancipator?

6. If you think of your own workplace and the means by which it is organized and managed, what would you say is the more important means of organization – labor, language, or power?

7. What is a good job or, more generally, what is quality of work? Evaluate the way your own computer usage is organized in light of these criteria.

Chapter 10

1. How would a positivist develop the requirements specification for a new computer system? What would be the basic strategy and the primary techniques? How would users be involved?

2. How would a hermeneutic develop the requirements specification for a new computer system? What would be the basic strategy? What would be the primary techniques? How would users be involved?

3. Is the spiral model as described in Boehm's paper (1988) an example of the hermeneutic circle?
4. Evaluate the options and strategies for professional development at your workplace.
5. Compare and contrast a positivistic and hermeneutic approach to the introduction of CASE tools.

Chapter 11

1. Examine the nursing case in chapter 7. Design a consulting strategy based on critical ideals. How would a strategy based on structuralism differ?
2. Does it make sense to develop a requirements specification for a new computer system based on critical or structuralist ideals?
3. Relate the taxonomy in Hirschheim & Klein (1989), which is based on two dichotomies – subjective-objective and order-conflict – to our taxonomy, which is based on the two dichotomies of people-computers and systems-change.
4. Discuss Floyd's paper (1987) in light of the contradictions between people and computers and between systems and change. Do you find her basic distinction between product and process clear and useful?
5. Give an example of a system (social or technical) that people don't like. Why don't they change it? What would it take to change it?

Chapter 12

1. Imagine yourself working in the computer department of a major bank. Characterize the most relevant perspectives of the bank from the point of view of developing and maintaining its computer-based information systems.
2. Consider the following four metaphors: computing machine, automaton, tool, and medium. For each metaphor, give examples of computer applications that are particularly well described by it. What does each metaphor emphasize as the

purpose and functionality of the system? Who are the users and what are their roles and interests?

3. Take a well-known computer application. Apply the four metaphors – computing machine, automaton, tool, and medium – to characterize and evaluate the system.

4. Think of a computer-based production control system in a factory, a transaction system in a major bank, an information system in a public library, and a monitoring and control system in an airport control tower. Discuss the relevance and interpretation of each of the four metaphors – computing machine, automaton, tool, and medium – in relation to each of these systems.

5. Use Adizes' typology – producer, administrator, entrepreneur, and integrator – to characterize yourself as a professional. What are your strengths and weaknesses?

6. Use Adizes' typology to characterize your project group. Is it a good team? What are its strengths and weaknesses?

Further Reading

The epigraphs are from Descartes' *Rules for the Direction of the Mind*, probably written in 1628, from Nietzsche's notes for the year 1888, and from Marcus Aurelius' (121–180) *Meditations*. The dictionary entries in chapters 1, 6, 9, and 11 are from *The American College Dictionary* (New York: Random House, 1959).

Chapter 1

Aiken is quoted in Ceruzzi (1986, p. 197) and McCarthy in McCorduck (1979, p. 93). Ceruzzi (1991) offers a fine discussion of what it was like when "computers were human." The already classical source for the concept of algorithm is Knuth (1973). The discussion of how to understand Church's and Turing's theses in Hofstadter (1979) is popular, and in Minsky (1967) a bit more technical. Naur (1972, 1982) has interesting things to say about the role of formalization in program development.

For more about the mechanistic world view, refer to a good encyclopedia, such as *The Dictionary of the History of Ideas* or *The Encyclopedia of Philosophy*, and look up Enlightenment, Nietzsche, Descartes, Leibniz, Mechanism. Weber (1958) is a classic on the process of modernization and rationality. The mechanistic world view is discussed and criticized by Mumford (1934). Hanson (1958) and Haugeland (1985) present fine discussions of how Galileo and Descartes used mathematics to represent mechanics. Bolter (1984) is good on the role of metaphors and the world view of computer science. Winograd & Flores (1986) forcefully attack the idea of knowledge as representation.

In this chapter and elsewhere we make extensive use of Mintzberg (1983) and Morgan (1986) whenever we discuss organizations.

Chapter 2

Langefors (1966) makes clear the distinction between data and information and discusses the use of information in organizations. Our discussion of the irrational use of information in organizations is drawn from Feldman & March (1981). The classical source for the notion of tacit knowledge is Polanyi (1958). Hofstadter (1979) puts to good use the Carroll dialogue, quoting it in full. On the problem of turning competence and knowledge into information and data, see Dreyfus & Dreyfus (1986) and Suchman (1987).

Plato's discussion of the danger of using books is found in his dialogue the *Phaedrus*. Speaking of that dialogue, our bicycle example is in several respects similar to Pirsig's (1974) motorcycle example. Our discussion of concepts as prototypes draws on work in cognitive science, as found in Rosch & Lloyd (1978). Notice that we are not saying that Marx and Freud were romantics, only that they relied rather heavily on romantic ideas. The first chapter of Taylor (1975) contains an excellent description of the romantic world view. The idea that reality is constructed is discussed at length in many of the contributions to Floyd et al. (1992), for example in Dahlbom (1992).

Chapter 3

We have borrowed the terms "hard systems thinking" and "soft systems thinking" from Checkland (1981), but we include in "hard systems thinking" what he calls "reductionism." The example of the publisher referred to in the text is inspired by Yourdon (1989), an influential proponent of hard systems thinking. Checkland (1981) gives an excellent overview of systems thinking in general and a presentation of his own very influential soft systems approach. There is more on this approach in Checkland & Scholes (1990). Flood & Jackson (1991) give a survey of systems methodologies in organizational problem solving.

Israel (1979) presents an introduction to dialectic thinking. Churchman (1971) and Hofstadter (1979) offer two very different discussions of the philosophical foundations of systems thinking.

The Wittgenstein quote is from Wittgenstein (1922), another source of inspiration for systems thinkers. We borrowed the analysis of schools as institutions for disciplining both students and teachers and the ideas about a pervasive and fundamental communication structure for control from Foucault's (1978) fascinating study of the history of prisons.

Chapter 4

The quote with which this chapter begins is from Simon (1981, pp.4–5). Software development methods, such as the general waterfall model, have long been dominated by systems construction ideas. In this chapter, Wirth (1971) plays the leading role as an early proponent of systems construction. See Naur (1972) for his comments on the 8-Queens problem. Parnas & Clements (1985) and Floyd (1987) offer good critical discussions of the systems construction approach. Newell & Simon (1972) present a more extensive discussion of the type of problem solving involved in systems construction.

Kuhn (1970), introducing the notion of paradigms in science, presents a fascinating theory of the progress of science. If you want to read more about Linnaeus and Darwin, you can do so in Mayr (1982).

Chapter 5

The idea of software development as evolution is not as old as the construction idea, but it has been around since the early 1970s. Naur (1972), Basili & Turner (1975), Floyd (1984), and Knuth

(1989) present influential discussions of this idea. A recent contribution to the prototyping approach is given by Budde et al. (1992). Parnas & Clements (1985) present a document-driven approach to fake a rational design process.

A presentation of the Cocomo model is found in Boehm (1981). The experiment comparing systems construction to evolution is found in Boehm, Gray, & Seewaldt (1984). Boehm (1988) and Mathiassen & Stage (1990) present attempts to formulate methods or principles combining these two approaches. A discussion of the spiral model and risk management is found in Boehm (1989). Naur's principles are in Naur (1972).

There is a comprehensive discussion of bounded rationality in Simon (1982). Uncertainty and complexity are relatively clear from an intuitive standpoint, but difficult to define formally. Simon (1981) presents an interesting discussion of complex systems, but wisely refrains to define them.

Knuth (1989) quotes, with approval, Hein's "grook" about erring. There are more grooks in Hein (1966–73). We have borrowed the turtle example from Dennett (1984).

Chapter 6

Our discussion and comparison of the construction, evolution, and intervention approaches owe a lot to Lanzara (1983), and our description of the intervention approach is partly based on Argyris & Schön (1978). The nursing case is borrowed from Bermann & Thoresen (1990).

The idea of "breakthrough by breakdown" is inspired by Madsen (1988). Ehn (1988) reports on several large intervention projects, notably Utopia, and discusses the theory behind them. Bjerknes, Ehn, & Kyng (1987) have several papers on the intervention approach. Mathiassen & Nielsen (1989) combine dialectic thinking with a soft systems approach. Winograd & Flores (1986) discuss design as intervention, Greenbaum & Kyng (1991) report on the practice of such design, Robey & Marcus (1984) have an interesting discussion of rituals in systems design, and Weinberg (1985, 1986) has good things to say about the role of being a consultant.

In chapter 4 of Taylor (1975) there is a good, short discussion of Hegel's dialectics.

Chapter 7

The three dimensions of quality – functionality, aesthetics and symbolism – are standard in design theory. Cross (1984) offers an influential collection of papers on design. Norman (1988) has written a rewarding little book on the quality of artifacts in general. Some of the essays, particularly "Evolution Versus Design," in his later book (1992) touch on many of the issues of our book. Pirsig (1974) became a cult book on quality. Schulmeyer & McManus (1987) and Vincent, Waters, & Sinclair (1988) are two standard texts on software quality assurance. Freedman & Weinberg (1972) have compiled an excellent handbook on how to practice quality control, with good comments on the psychology involved. Argyris & Schön (1978) present an interesting discussion of what it means to accept quality as a challenge to transcend. Sartre's discussion of essence and existence is found in Sartre (1948).

Chapter 8

The Bororo example is in Lévi-Strauss (1976, p. 286), a fascinating adventure story about his travels in Brazil, spiced up with comments on the philosophy of science. The distinction between engineers and tinkerers is in Lévi-Strauss (1966, ch. 1).

Argyris & Schön (1978) and Schein (1985) are good on how to understand and change organizational cultures. The maturity model in Humphrey (1988) has become an influential framework for software quality management. In Weinberg (1971) you will find the discussion of egoless programming. Chief programmer teams were introduced by Baker (1972), and cleanroom development by Selby, Basili, & Baker (1987).

Chapter 9

Braverman (1974) is the classic Marxist discussion of the "de-qualification of labor." Hirschheim & Klein (1989) discuss the different professional roles in systems development. Mumford (1983, 1987) introduced the sociotechnical approach in the information sciences. Read about radical political agents in Bjerknes, Ehn, & Kyng (1987) and in Ehn (1988).

The three knowledge interests are introduced in Habermas (1968). The idea of the ideal speech situation and the discussion of truthfulness, rightness, and sincerity is in Habermas (1984). The Winner quote and the reference to Robert Moses are from Winner's collection of essays (Winner 1986, pp. 10, 22f). Berman (1982) has a longer discussion of Moses and similar cases.

Chapter 10

Weinberg (1982) is quoted in DeMarco & Lister (1990, p. 4), a very useful collection of outstanding papers on software engineering. Keat & Urry (1985) give a good overview of contemporary philosophies of science and have a good bibliography. Burrell & Morgan (1979) have an influential discussion of different perspectives in research on organizations. Their typology is applied to systems development in Hirschheim & Klein (1989). Nurminen (1988) discusses different perspectives on computers and people and argues in favor of human-scale information systems. Weinberg (1986) has written a challenging book on how to become a technical leader, overcomming the limitations of a traditional computer perspective.

Dunlop and Kling (1991) and Forester (1989) are two collections of papers on the general theme of this chapter. Lewis Mumford (1934) is the classical source on technology and people.

Chapter 11

Again, Keat & Urry (1985) give a good introduction to both structuralism and critical theory. Our presentation of structuralism is mainly based on the work of Lévi-Strauss. You have to go to the younger Marx to find his theory of alienation. Habermas' attempts at a general theory of communication are in Habermas (1984). The theory's application within our field is advocated by Lyytinen & Klein (1985). Burrell & Morgan (1979) and Hirschheim & Klein (1989) are relevant sources for the discussion of the different approaches.

Boguslaw (1965) has written a classical book on utopian perspectives on computers and change, and Franz & Robey (1984), Kling (1980), Kling & Scacchi (1980) offer important contributions to our understanding of the use of computers in changing organizations.

Chapter 12

Schön (1983) presents an influential discussion of the role of reflection in professional practice. Nygaard & Sørgaard (1987), Nurminen (1988), and Andersen et al. (1990) are general discussions of the role of perspectives in systems development. Lakoff & Johnson (1980) present a great collection of metaphors. In Kling & Scacchi (1982) and Janlert (1987) there are more metaphors for computers. Adizes (1979) is the source for the typology of professional roles. Brooks (1987) is quoted in DeMarco & Lister (1990, p. 14).

References

Adizes, I. (1979) *How to Solve the Mismanagement Crisis: Diagnosis and Treatment of Management Problems.* Irvington: Adizes Institute.

Andersen, N. E. et al. (1990) *Professional Systems Development.* Hemel Hempstead, Hertfordshire: Prentice-Hall.

Argyris, C., & Schön, D. A. (1978) *Theory in Practice. Increasing Professional Effectiveness.* San Francisco: Jossey-Bass.

Baker, F. T. (1972) Chief Programmer Team Management of Production Programming. *IBM Systems Journal,* 11, no. 1, 131–49.

Basili, V. R., & Turner, A. J. (1975) Iterative Enhancement: A Practical Technique for Software Devlopment. *IEEE Transactions on Software Engineering,* SE-1, no. 4, 390–6 (December).

Berman, M. (1982) *All That Is Solid Melts into Air: The Experience of Modernity.* New York: Simon & Schuster.

Bermann, T., & Thoresen, K. (1990) Can Networks Make an Organization? In Bjerknes et al. (1990).

Bjerknes, G., Ehn, P., & Kyng, M., eds. (1987) *Computers and Democracy.* Aldershot, England: Avebury.

Bjerknes, G. et al., eds. (1990) *Organizational Competence in System Development.* Lund, Sweden: Studentlitteratur.

Boehm, B. W. (1981) *Software Engineering Economics.* Englewood Cliffs, N.J.: Prentice-Hall.

Boehm, B. W. (1988) A Spiral Model of Software Development and Enhancement. *Computer,* 21, 61–72 (May).

Boehm, B. W. (1989) *Software Risk Management, Tutorial.* Los Alamitos, Calif.: IEEE Computer Society Press.

Boehm, B. W., Gray, T. E., & Seewaldt, T. (1984) Prototyping vs. Specifying: A Multi-Project Experiment. *IEEE Transactions on Software Engineering*, SE-10, no. 3, 133–45 (May).

Boguslaw, R. (1965) *The New Utopians. A Study of System Design and Social Change*. Englewood Cliffs, N.J.: Prentice-Hall.

Bolter, J. D. (1984) *Turing's Man: Western Culture in the Computer Age*. Chapel Hill: University of North Carolina Press.

Braverman, H. (1974) *Labor and Monopoly Capital: The Degradation of Work in the Twentieth Century*. New York and London: Monthly Review Press.

Brooks, F. P. (1987) No Silver Bullet. *Computer*, 20, no. 4, 10–19 (April). Reprinted in DeMarco & Lister (1990).

Budde, R. et al. (1992) *Prototyping: An Approach to Evolutionary System Development*. Berlin: Springer-Verlag.

Burrell, G., & Morgan, G. (1979) *Sociological Paradigms and Organizational Analysis*. London: Heinemann.

Ceruzzi, P. E. (1986) An Unforeseen Revolution: Computers and Expectations, 1935–1985. In J. J. Corn, ed. *Imagining Tomorrow. History, Technology, and the American Future*. Cambridge, Mass.: MIT Press.

Ceruzzi, P. E. (1991) When Computers Were Human. *Annals of the History of Computing*, 13, 237–44.

Checkland, P. (1981) *Systems Thinking, Systems Practice*. Chichester, England: John Wiley.

Checkland, P., & Scholes, J. (1990) *Soft Systems Methodology in Action*. Chichester, England: John Wiley.

Churchman, C. W. (1971) *The Design of Inquiring Systems*. New York: Basic Books.

Cross, N., ed. (1984) *Developments in Design Methodologies*. Chichester, England: John Wiley.

Dahlbom, B. (1992) The Idea That Reality Is Socially Constructed. In Floyd et al. (1992).

DeMarco, T., & Lister, T., eds. (1990) *Software State-of-the-Art: Selected Papers*. New York: Dorset.

Dennett, D. C. (1984) *Elbow Room: The Varieties of Free Will Worth Wanting*. Cambridge, Mass.: Bradford Books/MIT Press and Oxford University Press.

Dreyfus, H. L., & Dreyfus, S. E. (1986) *Mind over Machine*. New York: Free Press.

Dunlop, C., & Kling, R., eds. (1991) *Computerization and Controversy. Value Conflicts and Social Choices.* New York: Academic Press.

Ehn, P. (1988) *Work-Oriented Design of Computer Artifacts.* Stockholm: Center for Working Life.

Feldman, M. S., & March, J. G. (1981) Information in Organizations as Signals and Symbol. *Administrative Science Quarterly,* 26, no. 2, 171–86 (June).

Flood, R. L., & Jackson, M. C. (1991) *Creative Problem Solving: Total Systems Intervention.* Chichester, England: John Wiley.

Floyd, C. (1984) A Systematic Look at Prototyping. In R. Budde et al., eds. *Approaches to Prototyping.* Berlin: Springer-Verlag.

Floyd, C. (1987) Outline of a Paradigm Change in Software Engineering. In Bjerknes et al. (1987).

Floyd, C., Züllighoven, H., Budde, R., & Keil-Slawik, R., eds. (1992) *Software Development and Reality Construction.* Berlin: Springer-Verlag.

Forester, T., ed. (1989) *Computers in the Human Context: Information Technology, Productivity, and People.* Cambridge, Mass.: MIT Press.

Foucault, M. (1978) *Discipline and Punish.* New York: Random House.

Franz, C. R., & Robey, D. (1984) An investigation of User-Led System Design: Rational and Political Perspectives. *Communications of the ACM,* 27, no. 12, 1202–209 (December).

Freedman, D. P., & Weinberg, G. M. (1972) *Handbook of Walkthroughs, Inspections and Technical Reviews.* Boston: Little, Brown.

Freeman, P., & Wasserman, A. I., eds. (1980) *Tutorial on Software Design Techniques.* Los Alamitos, Calif.: IEEE Computer Society Press.

Greenbaum, J., & Kyng, M., eds. (1991) *Design at Work.* Hillsdale, N.J.: Lawrence Erlbaum Associates.

Habermas, J. (1968) *Knowledge and Human Interests.* Boston: Beacon Press.

Habermas, J. (1984) *The Theory of Communicative Action. Vol. 1, Reason and the Rationalization of Society.* Boston: Beacon Press.

Hanson, N. R. (1958) *Patterns of Discovery.* Cambridge, England: Cambridge University Press.

Haugeland, J. (1985) *Artificial Intelligence: The Very Idea.* Cambridge, Mass.: MIT Press.

Hein, P. (1966–73) *Grooks,* vols. 1–5. Copenhagen: Borgens Forlag.

Hirschheim, R., & Klein, H. K. (1989) Four Paradigms of Information Systems Development. *Communications of the ACM*, 32, no. 10, 1199–216 (October).

Hofstadter, D. R. (1979) *Gödel, Escher, Bach: A Golden Eternal Braid.* New York: Basic Books.

Humphrey, W. S. (1988) Characterizing the Software Process: A Maturity Framework. *IEEE Software*, 5, no. 2, 73–79 (March). Reprinted in DeMarco & Lister (1990).

Israel, J. (1979) *The Dialectics of Language and the Language of Dialectics.* New York: Humanities Press.

Janlert, L.-E. (1987) The Computer as a Person. *Journal for the Theory of Social Behavior*, 17, 321–41.

Keat, R., & Urry, J. (1985) *Social Theory as Science*, 2d ed. London: Routledge & Kegan Paul.

Kling, R. (1980) Social Analysis of Computing: Theoretical Perspectives in recent Empirical Research. *Computing Surveys*, 12, no. 1, 61–100 (March).

Kling, R., & Scacchi, W. (1980) Computing as Social Action: The Social Dynamics of Computing in Complex Organizations. In M. C. Yovits, ed. *Advances in Computers*, 19, 249–372. New York: Academic Press.

Kling, R., & Scacchi, W. (1982) The Web of Computing: Computer Technology as Social Organization. In M. C. Yovits, ed. *Advances in Computers*, 21, 1–90. New York: Academic Press.

Knuth, D. E. (1973) *The Art of Computer Programming, Vol. 1, Fundamental Algorithms*, 2d ed. Reading, Mass.: Addison-Wesley.

Knuth, D. E. (1989) The Errors of TEX. *Software – Practice and Experience*, 19, no. 7, 607–85 (July). Reprinted in Floyd et al. (1992).

Kuhn, T. (1970) *The Structure of Scientific Revolutions*, 2d ed. Chicago: University of Chicago Press.

Lakoff, G., & Johnson, M. (1980) *Metaphors We Live By.* Chicago: University of Chicago Press.

Langefors, B. (1966) *Theoretical Analysis of Information Systems.* Lund, Sweden: Studentlitteratur.

Lanzara, G. F. (1983) The Design Process: Frames, Metaphors and Games. In U. Briefs et al., eds. *Systems Design For, With and By the Users.* Amsterdam: North-Holland.

Lévi-Strauss, C. (1966) *The Savage Mind.* London: Weidenfeld and Nicolson.

Lévi-Strauss, C. (1976) *Tristes Tropiques*. London: Penguin Books.

Lyytinen, K., & Klein, H. K. (1985) The Critical Theory of Jürgen Habermas as a Basis for a Theory of Information Systems. In E. Mumford et al., eds. *Research Methods in Information Systems*. Amsterdam: North-Holland.

Madsen, K. H. (1988) Breakthrough by Breakdown: Metaphors and Structured Domains. In H. K. Klein & K. Kumar, eds. *Information Systems Development for Human Progress in Organizations*. Amsterdam: North-Holland.

Mathiassen, L., & Stage, J. (1990) Complexity and Uncertainty in Software Design. In *Computer Systems and Software Engineering*. Los Alamitos, Calif.: IEEE Computer Society Press.

Mathiassen, L., & Nielsen, P. A. (1989) Soft Systems and Hard Contradictions. *Journal of Applied Systems Analysis*, 16, 75–88.

Mayr, E. (1982) *The Growth of Biological Thought*. Cambridge, Mass.: Harvard University Press.

McCorduck, P. (1979) *Machines Who Think. A Personal Inquiry into the History and Prospects of Artificial Intelligence*. New York: W. H. Freeman.

Minsky, M. (1967) *Computation: Finite and Infinite Machines*. Englewood Cliffs, N.J.: Prentice-Hall.

Mintzberg, H. (1983) *Structure in Fives: Designing Effective Organizations*. Englewood Cliffs, N.J.: Prentice-Hall.

Morgan, G. (1986) *Images of Organization*. Beverly Hills: Sage.

Mumford, E. (1983) *Designing Human Systems*. Manchester Business School, Manchester, England.

Mumford, E. (1987) Sociotechnical Systems Design. Evolving Theory and Practice. In Bjerknes et al. (1987).

Mumford, L. (1934) *Technics and Civilization*. New York: Harcourt Brace Jovanovich.

Naur, P. (1972) An Experiment on Program Development. *BIT*, 12, no. 3, 347–65. Reprinted in Naur (1992).

Naur, P. (1982) Formalization in Program Development. *BIT*, 22, no. 4, 437–53. Reprinted in Naur (1992).

Naur, P. (1985) Intuition in Software Development. In H. Ehrig et al., eds. *Formal Methods and Software Development*. Berlin: Springer-Verlag. Reprinted in Naur (1992).

Naur, P. (1992) *Computing: A Human Activity*. Reading, Mass.: Addison-Wesley.

Newell, A., & Simon, H. A. (1972) *Human Problem Solving.* Englewood Cliffs, N.J.: Prentice-Hall.

Norman, D. A. (1988) *The Psychology of Everyday Things.* New York: Basic Books. Reprinted as *The Design of Everyday Things.*

Norman, D. A. (1992) *Turn Signals Are the Facial Expressions of Automobiles.* Reading, Mass.: Addison-Wesley.

Nurminen, M. I. (1988) *People or Computers: Three Ways of Looking at Information Systems.* Lund, Sweden: Studentlitteratur.

Nygaard, K., & Sørgaard, P. (1987) The Perspective Concept in Informatics. In Bjerknes et al. (1987).

Parnas, D. L., & Clements, P. C. (1985) A Rational Design Process: How and Why to Fake It. In H. Ehrig et al., eds. *Formal Methods and Software Development.* Berlin: Springer-Verlag. Reprinted in DeMarco & Lister (1990).

Pirsig, R. M. (1974) *Zen and the Art of Motorcycle Maintenance.* London: The Bodley Head.

Polanyi, M. (1958) *Personal Knowledge.* London: Routledge & Kegan Paul.

Robey, D., & Marcus, M. L. (1984) Rituals in Information System Design. *MIS Quarterly,* 8, no. 1, 5–15 (March).

Rosch, E., & Lloyd, B. B., eds. (1978) *Cognition and Categorization.* Hillsdale, N.J.: Lawrence Erlbaum Associates.

Sartre, J.-P. (1948) *Existentialism and Humanism.* London: Methuen.

Schein, E. K. (1985) *Organizational Culture and Leadership. A Dynamic View.* San Francisco: Jossey-Bass.

Schön, D. (1983) The Reflective Practitioner. *How Professionals Think in Action.* New York: Basic Books.

Schulmeyer, G. G., & McManus, J. I. (1987) *Handbook of Software Quality Assurance.* New York: Van Nostrand Reinhold.

Selby, R. W., Basili, V. R., & Baker, F. T. (1987) Cleanroom Software Development: An Empirical Evaluation. *IEEE Transactions on Software Engineering,* SE-13, no. 9, 1027–37 (September). Reprinted in DeMarco & Lister (1990).

Simon, H. A. (1981) *The Sciences of the Artificial,* 2d ed. Cambridge, Mass.: The MIT Press.

Simon, H. A. (1982) *Models of Bounded Rationality: Behavioral Economics and Business Organization.* Cambridge, Mass.: The MIT Press.

Strunk, W., & White, E. B. (1979) *The Elements of Style,* 3d ed. New York: Macmillan.

Suchman, L. (1987) *Plans and Situated Actions*. Cambridge, England: Cambridge University Press.

Taylor, C. (1975) *Hegel.* Cambridge, England: Cambridge University Press.

Vincent, J., Waters, A., & Sinclair, J., (1988) *Software Quality Assurance.* Englewood Cliffs, N.J.: Prentice-Hall.

Weber, M. (1958) *The Protestant Ethic and the Spirit of Capitalism.* New York: Charles Scribner's Sons.

Weinberg, G. M. (1971) *The Psychology of Computer Programming.* New York: Van Nostrand Reinhold.

Weinberg, G. M. (1982) Overstructured Management of Software Engineering. In *Proceedings of the Sixth International Conference on Software Engineering.* Tokyo, Japan. Reprinted in DeMarco & Lister (1990).

Weinberg, G. M. (1985) *The Secrets of Consulting. A Guide to Getting and Giving Advice Successfully.* New York: Dorset House.

Weinberg, G. M. (1986) *Becoming a Technical Leader: An Organic Problem Solving Approach.* New York: Dorset House.

Weston, A. (1987) *A Rulebook for Arguments.* Indianapolis and Cambridge, England: Hackett Publishing.

Winner, L. (1986) *The Whale and the Reactor.* Chicago: University of Chicago Press.

Winograd, T., & Flores, F. (1986) *Understanding Computers and Cognition: A New Foundation for Design.* Norwood, N.Y.: Ablex.

Wirth, N. (1971) Program Development by Stepwise Refinement. *Communications of the ACM,* 14, no. 4, 221–27 (April). Reprinted in Freeman & Wasserman (1980).

Wittgenstein, L. (1922) *Tractatus Logico-Philosophicus.* London: Routledge & Kegan Paul.

Yourdon, E. (1989) *Modern Structured Analysis.* Englewood Cliffs, N.J.: Prentice-Hall.

Index

Printed and bound by CPI Group (UK) Ltd, Croydon, CR0 4YY

27/10/2024

14580369-0004